This Is What Leaders Do

Seven Essentials to Inspire and Empower Your Team

This Is What Leaders Do

Seven Essentials to Inspire and Empower Your Team

Russell E. Justice

KeyPress Publishing

KeyPress Publishing
www.keypresspublishing.com

KeyPress Publishing

This book is a work of nonfiction. Unless otherwise noted, the author and the publisher make no explicit guarantees as to the accuracy of the information contained in this book and in some cases, names of people and places have been altered to protect their privacy.

"Mr. Cabby" artwork designed by John W. Cook. Used by permission.

Story cloth used with permission by Calvary Road Ministries, https://calvaryroadministries.org

THE HOLY BIBLE, NEW INTERNATIONAL VERSION®, NIV® Copyright © 1973, 1978, 1984, 2011 by Biblica, Inc.® Used by permission. All rights reserved worldwide.

Author: Russell E. Justice

This Is What Leaders Do: Seven Essentials to Inspire and Empower Your Team

Published by: KeyPress Publishing
Science Adviser: Thomas Freeman
Brand Integrity and Design: Jana Burtner
Production Manager: Adele Hall
Editors: Gail Snyder, Ashley Johnson, Stefanie Carr, and Kelly Lee

ISBN: 979-8-9922514-2-5 (Paperback); 979-8-9922514-1-8 (Hardcover)

Library of Congress Control Number: 2025930395
Published in Melbourne, Florida

Distributed by:
ABA Technologies, Inc.
930 South Harbor City Blvd, Suite 402
Melbourne, FL 32901

www.abatechnologies.com

KeyPress Publishing books are available at a special discount for bulk purchases by corporations, institutions, and other organizations. For more information, please email keypress@abatechnologies.com.

Praise for
This Is What Leaders Do

"I believe the world needs more effective leaders who know how to forge new paths, marked with deep values for people ... This book is both the *what* leaders should do as well as the *how* leaders should do it."

—Dr. Victor Dingus, PE, FELLOW, DMin
Innovation Coach

"The principles gained from my close association with Russell Justice influenced every phase of my life including my family, marriage, faith, friendships, and the parenting of my children. All phases of life are processes, and all can be improved using the principles included in this book."

—Earnest W. (Earnie) Deavenport
Retired Chairman and CEO, Eastman Chemical Company

"This Accelerated Continuous Improvement as taught to us by Russell was the most fun and satisfying experience in my career. This book lays the process out beautifully, and the many case files he spells out help you understand it and become confident that you can do it too."

—Larry Speight
Retired Vice President/General Manager, Honeywell
Space Systems

"The stories in this book will stick with you like molasses on a biscuit with their clear steps that team members will remember and use again and again. Bonus: The author's engaging storytelling style makes the learnings long-lasting and fun."

—Janis Allen
Owner, Performance Leadership Consulting

"I personally learned these principles from Russell over 30 years ago and I still use them today in my consulting practice. Best of all, when the outstanding results come and the success stories are told, the resounding comments are—'That was FUN!'"

—Eileen Flynn
President, Legacy Teams, LLC

"Stories are a powerful learning tool because they are simple and memorable. Russell is a master storyteller with the case studies and examples in this book. We learn best by doing, as Russell teaches us. Read this book and then start doing Russell's seven steps. You'll be glad you did."

—Michael McCarthy
Author of *Sustain Your Gains*

"We had tried many continuous improvement programs over the years, but it wasn't until we were introduced to Russell Justice's ACI program that we learned meaningful continuous improvement. This book embodies his down-to-earth approach and simple but effective methods that make it the one we follow today."

— John Harlow
Retired Distribution Manager, American Greetings

Dedication

This book is dedicated to Aubrey Daniels, my lifelong friend and mentor. Without Aubrey's influence, there would be no book to be written. What I like to say most about him is, "I don't know who I would be if I had not met him." I never met a better man and never had a better friend. Aubrey profoundly changed my life—my career, my marriage, my parenting and grandparenting, and my ministry. I have told literally thousands of people all over the world about this incredible Southern gentleman named Aubrey Daniels and how he touched my life.

For those unfamiliar with Aubrey, he founded Aubrey Daniels International (ADI) and is the world's leading and internationally recognized authority on applying behavioral science in the workplace, including on leadership and management issues. He is also the author of best-selling books on the subject and has received numerous awards for his work in the field of behavior analysis. It has been my blessing to see thousands of people's lives and work improved because of his teachings. As an extension, because of what he taught me in workshops, over dinners, on the phone, in his office, and as we traveled, I have been able to extend that impact. Aubrey poured into me the science of behavior and the maxims of living like a professional and a gentleman.

His zeal is infectious. His character is a role model for anyone who is privileged to spend time with him. His charisma with folks from the front lines to the office of the CEO is impressive. His gravitas walking into a room of executives is amazing. He is the master storyteller, with stories that are spellbinding and unforgettable. He sets the standard for how to conduct yourself, how to be friendly, approachable, and kind with everyone who comes across your path. He sets the standard in how to dress as a professional. He taught me that clothes don't make the man; they reveal the man. Once when I was admiring his tie, without hesitation, he just undid it and handed it to me. It is a gift that I treasure still today and a tradition that I sometimes follow when someone admires my jacket, tie, or cap.

Our relationship began with what was (to this day) *the* best learning experience I have ever had—the 2-week Performance Management Training that he led in Tucker, an Atlanta suburb, June 13-24, 1983. We instantly hit it off because of my hunger to know more and his unequaled expertise about applied behavior analysis (ABA). I would stay long after class (and he indulged me) to ask questions and talk about the principles and approaches. He taught me about collecting "gems," which has been a life-shaping practice and gave me the Gem of the Decade for the 1970s: "Behavior is a function of consequences."

We experimented with communicating using this thing called AOL Instant Messaging in its early days. We traveled the United States together to company locations, for professional meetings where he agreed to speak when I asked him, and to ABA national conventions where his arrival was like, "Elvis is in the building!" I had the privilege of introducing his teachings in over 20 countries. (Yes, the principles and concepts of ABA do work in every culture, just like gravity works.) As the years went by, we came to agree that we were no longer sure if the stories we told "belonged" to him or to me. Attending Braves baseball games together became a tradition. His teachings played a major part in helping Eastman Chemical Company and Pal's Sudden Service win the Malcolm Baldridge National Quality Award and in helping Kodak Australia win the Australian Quality Prize.

At a quarterly RealLife Men's meeting, we had a great breakfast and then a video about mentoring. One of the tabletop questions we discussed after the video was, "Who would you say are the men who have had major influence in making you the man you are today?" Howard Hendricks and Aubrey Daniels were two men I listed as changing my life forever. Howard was talked about on the video we watched. Then ... as I had been called up front to recognize my 70th birthday coming up the next weekend ... from the back, *Aubrey Daniels* comes walking up! I was completely thrilled! Aubrey had come to participate in my surprise birthday celebration. His making the trip from Atlanta to Charlotte for my birthday made a special day even more special and unforgettable.

Thank you, Aubrey! I'm thankful God knit you together and gifted you with the passion to learn and to share it with others like me. Thanks for being my teacher, partner in business excellence, mentor, guru, coach, role model, accountability partner, and friend of a lifetime.

"The Lord bless you and keep you; the Lord make his face shine on you and be gracious to you; the Lord turn his face toward you and give you peace." (Numbers 6:24-26)

Contents

Foreword

I was looking out the window of the manufacturing facility I was responsible for managing. Jack was walking from the parking lot toward the employee entrance. His gait was slow, and his stature seemed to diminish under the weight of the impending day's battle to make customers and employees happy. It made me sad to think that Jack's job was something to be suffered through.

Several weeks before, my new HR manager Tammy asked me how we were going to lead improvement. The question caught me off guard, as we had a structured approach utilizing improvement teams. I thought I had sufficiently explained our approach but did so again anyway.

After I finished she said, "Right, but how are we going to lead improvement?" She said, "I have someone you need to meet—Russell Justice."

A short time later, the day came for the meeting with Russell. It was a very busy day, and several problems had me consider canceling the meeting. I decided to keep the meeting, and as a result, did not miss one of the most important days in my life—the day I met Russell.

Russell had a full beard, wore glasses, and had a thoughtful look about him. Absent were the pressed suit and fancy brochures. It didn't feel like a sales call but rather a response to a request for help. I didn't realize that when Tammy called him to set up a meeting, she said I needed his help.

Up to that point, I had had some success doing improvement with Theory of Constraints, Lean methods, Six Sigma, and Total Quality Management tools. But doing improvement was a battle ... getting buy-in; vying for needed resources; and if solutions were found—keeping them in place.

I could sense something different about Russell, but I couldn't put my finger on it. His part of the discussion was absent the heavy-handed, heady, and pushy approach of so many others. Though I didn't fully understand what I was getting into, I decided to start a relationship with Russell.

He changed my life.

Russell taught me What Leaders Do. I started leading improvement instead of doing improvement. He taught me to combine the Total Quality Management tools I had been using with the science of human behavior and the power of positive reinforcement. What we did together reads like a bold book title and the hyperbole so many use to sell books. Difference is, we did it. All modes of performance improved—we set new records in safety, quality, delivery, cost, and engagement.

Aubrey Daniels often said that Russell was and is the best he has ever seen at leading behavioral improvement. Aubrey said that Russell should write a book. He did, and here it is.

I encourage you to read the book, meet Russell, and change your life.

—Jay S. Roths
R Enterprises, LLC

Preface

The first responsibility of a leader is to define reality.
The last is to say thank you. In between the two,
the leader must become a servant and a debtor.

—Max De Pree, *Leadership Is an Art*

I was in Greeneville, Tennessee, in late September for the first meeting of a newly forming manufacturers' council. It was an honor to speak to this group representing the leadership of more than 15 local manufacturing organizations. I had 1 hour at lunchtime to share on the topic: accelerating continuous improvement.

As I prepared for the presentation, I kept adding the words "This is what leaders do" to my notes. Practicing the night before and the next morning, I realized that these words emphasized the thought that the most important thing I could share was "what leaders do" to accelerate continuous improvement. Because knowing that accelerated improvement is critical to the future of the organization is one thing; knowing *how* to do it is another matter completely.

After the presentation, I asked the leaders to each turn in their "gem of the presentation"—the most important principle, concept, point, statement, or challenge they heard during the presentation. Below is a summary of those "gems" from my presentation This Is What Leaders Do.

Leaders:

- Focus the organization on the vital few (one or two) critical business issues.
- Create alignment of efforts by linking teams and individuals in to the focus. Create a magnetic field of "teamwork flux," and orchestrate improvement.
- Bring out the discretionary effort in associates by creating "want to" systems instead of "have to" systems.

- Fulfill two obligations when asking people to do something. Observe when it is done, and acknowledge it.

- Stick to the focus by stiff-arming distractions.

- Define what's on the y-axis.

- Make performance visible by using scoreboards to turn measurement into feedback that is timely and specific enough to facilitate change.

- Teach all associates about compensating, correcting, and preventing actions and about root cause.

- Charge all associates with two jobs: (1) Do the job the best-known way today (best practices), and (2) find a better way to do it tomorrow (innovation).

- Celebrate improvements and reinforce the behaviors leading to those improvements.

- Know what to reinforce, who to reinforce, when to reinforce, and how to reinforce.

- Know how to "dance in the end zone" by following the four steps of reinforcement:
  - What did we do?
  - Why is it important?
  - How did we do it?
  - Enjoy the accomplishment.

This book, based on these gems, defines the elements of this leadership approach and provides an abundance of application examples through case files and case studies. These examples will point you in the right direction and give you tips to keep you going on your journey to performance excellence.

Renowned American economist Dr. Edwards Deming's 14th point for management states, "Put everybody in the company to work to accomplish the transformation" (Deming, 2018, p. 23). Deming, known as the Father of Quality, recognized that outstanding leadership is required to "make it happen." He understood that just knowing the right thing to do (points 1–13) was not enough. A conscious, deliberate, intentional effort was required to make it happen. When I asked him once how to go about

xvi

putting everyone in the organization to work, he said I would have to talk to someone else about this. After hearing me explain the approach shared in this book to him on stage once, he said, "I can find nothing wrong with it. You have to start somewhere" (personal communication, 1990). I consider this a high compliment from him. This book addresses the necessary leadership excellence for world-class performance. These principles, concepts, and examples will help you turn your plans for improvement into a reality.

Writing this book is not just a priority for me; it is an essential. Priorities are chosen; essentials are assigned. The single greatest "why" for writing this book is the resolute encouragement (push) from my lifelong mentor Dr. Aubrey Daniels to "Write the book!" Dr. Deming said, "Those who have been fortunate to have these experiences have an obligation to share them with others." In addition, I have received a stream of encouragement from clients and colleagues throughout the years to capture these principles and examples of performance excellence.

Upon turning 75, my life theme became "number my days," prompted by Psalm 90:12: "Teach us to number our days, that we may gain a heart of wisdom" and by the Latin phrase "memento mori" (remember death). At age 75, I also wrote down my hankerings and yearnings—one of which was "Write the book!" Lastly, a gem that has always spoken to me is, "When an old person dies, it's like a library burning down." With that somewhat somber thought in mind, I am sharing some of my library while I'm still here!

I've been blessed beyond measure to spend a lifetime working with organizations around the globe to bring about improvement. I'm excited about sharing these examples, case studies, and case files with you. I'll spend some time talking about *what* to do, but there will be much emphasis on examples. I believe providing examples is a superior way to help you. I'm confident that as you see these examples, they will trigger ideas of your own of *how* you can accelerate continuous improvement in your world. So, saddle up your horses and let the great adventure begin!

Acknowledgments

There is no way possible to acknowledge all those who have contributed to the journey that led to this book. But I would be remiss if I didn't call out many who made significant impact.

Acknowledgments should begin with the one person absolutely essential in this book coming to life. My wife for 54 years, Debby. She has given me the time and space needed to do the writing and has been a never-ending source of encouragement. She has been patient with me when I was too often distracted with clients, projects, and this book. She has traveled around the globe with me and kept the home fires burning when I was away. Thank you so much, my sweetheart.

The principles and concepts and the examples shared here are an outcome of working with many incredible folks. I enjoyed not only working together but amazing days of friendship as well. Among those people:

- **Organization Leaders:** Toy Reid, Earnie Deavenport, Wiley Bourne (Eastman Chemical Company); George Trabue, Jack Lowe (Eastman Chemicals Marketing); Bob Hart, Bill Garwood, Charlie Bailey (Tennessee Eastman); John Beckler (Carolina Eastman); Jay Roths (Angus Palm & J Enterprises); Scott Crawford (American Greetings); Larry Speight (Honeywell Space); Mike Truscio (Honeywell ATS); Steve Szilagyi (Lowe's Home Improvement); David Hart (Tennessee Rehabilitation in Corrections); Pal Barger, Thom Crosby (Pal's Sudden Service); Geoff Bailey (Arby's); Rick Rose, Alex Anderson (Barter Theatre); David and Skippy Swanger (Alabama Sheriff's Boys Ranch).

- **Continuous Improvement Associates/Partners** who worked alongside me in workshops and seminars to design and implement improvements and also taught, coached, and mentored me: Victor Dingus, Bob Gerwig, Al Robbins, David McClaskey, Keith Scott, Ivey Redmon, Jim Fuller, Janis Allen, Alex Zalesky.

- **On-Site Continuous Improvement Coordinators:** Eileen Flynn, Marsha Beam, Rick Creecy, Ian MacDonald, Susan Cunningham, Ramone Alvarado, Anne-Marie Leslie.

- **Advocates for ACI:** Bob Settles (American Greetings); Tammy Gracia (Angus Palm); Tom Holliday (RPM Pizza); Tom Carter (Alcoa); Butch Kinsey (Lowe's); Neil Farmer (A&W Canada).

- **Leaders on the Front Lines** who took the concepts, principles, and approach and made it work to improve their work areas and performance: Scott Felts (American Greetings); Leigh Dyson (Alcoa Australia); Roger Clark (TRICOR); Ross Baxter (Kodak Australasia); Mario Cangas (Eastman Chemicals Argentina); Jair DeBritto (Eastman Chemicals Brazil).

- And to all **those who attended** my workshops, seminars, and conference and said, "write the book."

- Lastly, I would like to extend special thanks to Ashley Johnson and Gail Snyder for their work in **editing the book**. Their eyes for details, meticulous proofreading, keen insight to the message of the book, valuable suggestions, and dedication to excellence shaped the manuscript significantly and enhanced the quality of this work.

Introduction

A question I have found very useful when stuck in a situation like an overbooked flight, a lost hotel reservation, or no appointments available at the doctor's office is, "What would you do if you were in my situation?" A similar far-reaching question I have discovered in my professional career is, "Can you give me an example of that?" This book is centered around those questions and the answers given through a collection of case files and case studies. It includes examples of what leaders and organizations have done to bring about significant, value-adding results in vital business areas through creativity, imagination, discipline, and hard work. The examples are presented in the context of seven essential elements of leadership required for Accelerated Continuous Improvement (ACI).

As the saying goes, "You can't teach what you have not experienced any more than you can come from where you have not been." What you will read here is based on 60 years of job experience: picking cotton in the fields, selling vegetables from the back of a truck, working in Advanced Systems at NASA, 27 years at Eastman Chemical Company, and consulting and coaching with over 80 organizations around the globe. In this book, you will see dozens of examples from those 60 years of the actions taken to produce phenomenal results in organizations from many sectors of business, large, medium, and small.

This book was not written by starting with an outline or a list of the chapters. It was written starting with the outcomes—what readers will be expected to do differently after reading and studying the book. We might be tempted to say how it will change your *thinking*, but that would not be sufficient. What is required is a book that will change the *behaviors* of leaders to produce ACI in the vital areas of their business, and to do that in an uncomplicated way, even to the point that leadership becomes simplified and easier, not more complex. People tend to make leadership far too hard; it's just not that hard. For that reason, this book has two purposes: (1) to give you a template that, when utilized, will result in ACI and (2) to simplify your life by giving you the vital few components of effective leadership and the examples to help you move forward.

The concepts and examples you read about and study here will enable you to accomplish the following:

- Focus your organization on *the* vital few issues for your business.
- Align and mobilize the efforts of your workforce toward the focus.
- Go beyond just projects to enterprise-wide improvement.
- Tap into everyone's discretionary effort (the difference in "have to" and "want to").
- Get unstuck when tackling a problem and at a dead end.
- Trigger innovative and creative ideas that directly apply to your organization.
- Experience quality *of* management.

At this point, you should stop reading for a moment and make sure you have the following:

- A tool for marking up this book: Underline, star, make notes in the margin. Make the book more valuable to you by adding your thoughts and notes.
- A place to write down action ideas that will come to you about how to improve your homelife and work life.

Navigating the Book

Throughout the book, you will find symbols to help you better navigate the book. These will serve as guideposts for important principles, concepts, and ideas. When you encounter one of these, take a breath, slow down, read more carefully, and pay more attention.

You will find symbols for the following:

CASE FILE

An example that illustrates one of the seven key elements for leadership excellence.

CASE STUDY

An example that illustrates multiple elements for leadership excellence.

KEY POINT

Tags a concept, principle, or idea to be emphasized.

TIPS

Additional material to help with understanding and implementing one of the elements.

THE STORY

The telling of an experience.

How to Get the Most Out of This Book

- Get a copy for each member of your team.
- Schedule a weekly lunchtime meeting to review the book.
- Ask team members to read one chapter per week and identify the most important point in that chapter (the gem).
- As the team members arrive at the weekly meeting, have a flip chart, white board, or screen labeled "Gems From the Chapter."
- Before taking a seat, each team member writes their gem on the flip chart, *then* picks up their lunch (earned by bringing a chapter gem).
- During lunch, go around the room and let each team member share their gem and elaborate on why they chose it.
- Type up the gems and distribute them to all who attended.

Alternate approach: Have each team member write a "test question" and the answer from the chapter. (You learn more by writing the questions than by taking the test.) Let the team vote on the top 3–5 questions from each chapter.

Note: For a virtual meeting, make the needed adjustments to this process.

Overview of the Approach

Many organizations have continuous improvement as one of their core guiding principles. However, the question is **not**, "Are you continually improving?" The question is, "Are you continually improving **faster** than your competition?" Many organizations are continually improving yet still not nearly reaching their full potential, and some are even going out of business. The reason? They are not improving fast enough.

Stated more precisely—the issue for all businesses is the **rate** of continuous improvement. Your organization is faced with the challenge of **accelerating** its rate of improvement.

The purpose of this book is to share with you a methodology, a philosophy, an approach to "change the slope of the improvement line"—to accelerate improvement. I will show you how to do that by sharing with you the principles and concepts of ACI and an array of examples, case files, and case studies. ACI is a systematic, data-based, and conscious application of the science of applied behavior analysis (ABA) and the Total Quality Management (TQM) approach to speed up (accelerate) and sustain performance improvements.

You are not learning to *do* improvement (we have problem-solving approaches for that), but, more so, you are learning how to *lead* improvement by creating a situation where people care about performance and want to improve. Think of this approach as a track (like a railroad track)—an improvement track—for your leadership and the efforts of your organization to run on. The track provides direction and boundaries.

Perhaps an even better analogy would be a trellis for grapevines or a tomato stake. The purpose of the trellis or stake is to provide "discipline" for the plants, not to restrict them in any way, but to help them grow. The purpose is to keep them off the ground, where they would surely rot, and to help them reach toward the sunlight, catch the rain, and thrive.

I'm assuming that you are continually improving, like we see on the line below the triangle in Figure 0.1. The arrow above the triangle represents the accelerated improvement achieved by using the principles and concepts of ACI. The triangular area represents the added value from the accelerated improvement and the competitive advantage that you will gain.

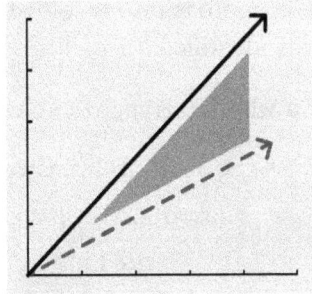

Figure 0.1. The current rate of improvement must be accelerated to achieve competitive advantage.

We are not looking for the typical 5% improvement. We're looking for improvement where the scale on the graph has to be changed because it no longer makes sense. Take this project in Texas to reduce errors in the bills of lading: Starting with a graph displaying the percentage of bills correct, with a scale from 85% to 100%, with typical weekly results around 90%, the weekly accuracy numbers climbed through the goals of 95% and 97% over the first 6 months. Before year's end, the graph was consistently showing 100%—so much so that it really was not a graph anymore, just a line. The supervisor had written a note on the graph next to a string of weeks at 100% saying, "No end in sight!" (In Chapter 6, when we discuss goals, we'll talk about what to do next ... after perfection.)

I have made the assertion many times and to thousands of people in workshops and presentations that ACI is the best-known approach in the world today to accelerate improvement. I have asked them, after hearing the approach, if they know a better one to please share it with me. So far, there have been no takers.

The reason this approach is the best is because it fully integrates the concepts, principles, and methods of TQM and ABA (Figure 0.2). This is not just tacking one onto the end of the other but combining them into a compound, where the

Figure 0.2. ABA and TQM partnership: The best-known method in the world today to accelerate improvement.

two approaches are joined in such a way that they cannot be individually identified.

So, what is this approach? Let's take a look at it. It embodies seven elements:

- Focus/Pinpoint: Create a unifying theme.
- Kickoff: Conduct a kickoff to engage the workforce.
- Translate & Link In: Help teams translate & link in.
- Management-Action Plan: Develop & carry out a management-action plan.
- Improve Processes: Teams work to improve processes.
- Measure & Feedback: Measure progress & provide feedback.
- Reinforce & Celebrate: Reinforce behaviors & celebrate results.

For each of these elements, let's look at the concepts and principles, and of course, real-world examples of their application.

"If you learn only methods, you'll be tied to your methods. But if you learn principles, you can devise your own methods."

—Widely attributed to Ralph Waldo Emerson

CHAPTER 1

Focus/Pinpoint—Create a Unifying Theme

Let's Get Started!

Let's begin with focus, what is often called "pinpointing." I like the word pinpointing. It sounds just like what it means: zeroing in—to a laser focus.

It has been my experience that the single greatest cause of failure (by that, I mean not achieving all that is possible) in organizations is the lack of focus. Many leaders I have worked with simply refuse to focus. They believe that they can't focus—they must work on everything. They are wrong.

In my more than 50 years of helping organizations to improve, I have found that the single greatest cause of "failure" is lack of focus.

We are not abandoning the many aspects of a business or organization that need attention but zeroing in on one (or maybe two) key result areas. These other aspects of business will still be improved as problems arise and we resolve them. They will be addressed reactively while we are addressing the pinpoint proactively.

For our pinpoint, we will select the one area of business that will most impact the future of our organization. Having identified that area, we will drive it (proactively) toward improvement—accelerated improvement.

We will come back to this discussion, but first let's look at some examples of pinpoints, since an example is often the best way to explain an idea or concept and because this book is dedicated to providing you with helpful examples. Each of these pinpoints has a story. We will cover many stories later, but for now, we will look only at the pinpoint (and in some cases the theme). As you look at these, they will spur your thinking about what the pinpoint might be for your organization. Begin taking notes now on your ideas. Use the margins in the book for your notes and ideas, or start a notepad now.

CASE FILE

Refining Uptime—Bauxite Mining and Alumina Refining

With a competitive market and an operation that is very corrosive on the equipment, a few percent increase in uptime can make a huge difference in production and provide a competitive advantage.

CASE FILE

Fill the House—Regional Theater

Reservations, seating, parking, refreshments, and restaurants nearby (all impacting the pinpoint) were possible pinpoints for the theater, but it was decided that *the* critical factor, the pinpoint, for surviving was to fill the house (seats).

CASE FILE

Pizza Remakes

Remakes caused by dropping pizzas, over/undercooking them, making ingredient mistakes and order errors, and receiving changes to orders—all must be eliminated since they decrease resource utilization of materials, equipment, and people and impact delivery time.

CASE FILE

High & Tight—Shipping Cube

When the trucks leaving the warehouse for the retail store are packed high and tight, it decreases damage and reduces freight. However, high and tight requires extra attention to detail.

CASE FILE

Get Out and Stay Out (GOSO)

What should be the overarching aim for a statewide prison rehabilitation program? It is to help offenders to, first of all, get out by fulfilling their sentence with good behavior. They also need to get out with some new skills and resources that will help them stay out.

CASE FILE

Make International Business Easy (MIBE)

With the most potential for future growth in international markets and the lowest customer satisfaction scores, the question must be asked: "What are all the things we do that make life/business more difficult for our customers in other countries?" This includes areas like customs documents, invoice errors, and exchange-rate problems. The challenge became to make international business easy.

CASE FILE

Saving Trees–Paper Waste

In the business of making greeting cards and wrapping paper (or any other paper product), wasted paper is a huge issue. The theme became to not just talk about "waste," but to talk about "saving trees" and expressing waste as the equivalent number of trees cut down today because of our waste. Over time, the size of the forest that could be saved was projected.

Theme ideas include the use of axes, chain saws, lumberjack clothing, lumber, and video clips from *Swamp Loggers*, *Ax Men*, and *Extreme Loggers*.

CASE FILE

Got More Milk

A theme for overall efficiency on a dairy farm focused on the amount of milk per cow per day. And believe me, you would not imagine all the factors involved!

CASE FILE

On Track—Machine Uptime—Plant Maintenance

With over 500 buildings on-site and thousands of pieces of equipment, where to start? Start with the 20 most critical/bottleneck pieces of equipment and keeping them "on track" and out of the pits. This included two objectives: (1) When a machine is down, get it back on track as fast as possible, and (2) get it back on the track so as not to come back into the pits.

CASE FILE

Railcar Hold Times

When managing a fleet of expensive hopper cars to shuttle plastic pellets to the company where they were blown into bottles, it was critical to keep the hopper cars moving by reducing the unloading/turnaround time.

CASE FILE

Shorten the Shorts

When picking items from the bins to ship to a customer and one of the picks is missing, you have a "short." Challenge: "Shorten the Shorts."

CASE FILE

Time to Fill/Time to Proficiency

For Human Resources (HR), this is the time it takes to fill a vacancy and then, once hired, the time for the new employee to reach proficiency.

CASE FILE

All Pro

These are the employees who have become "**PRO**ficient" in all the jobs in their department.

THE STORY

Stop the Stops

I'm in Kentucky for my first visit to the greeting card factory there. The facility is huge, stretching for hundreds of yards and filled with machines to print, cut, fold, and package cards. As we begin to walk through the plant, I notice that a number of the machines are not running. I ask why. Are they not needed for production to meet demand? As I ask about machine after machine that is idle, I'm told that this one is down because of a mechanical problem, the next one because of ink problems, another one for a paper jam or static in the paper or color balance or a plate change, and one because the operator is out sick today.

By the time we got to the other end of the factory, it seemed clear to me what the pinpoint was—machine uptime—or stated more as a rally-able theme—"stop the stops." My tour guide points out dozens of other challenges, but I'm thinking stop the stops is the place to start.

With traffic-style lights placed at each machine, the procedure would be that when the machine is running the traffic light would be green, but any time the machine goes down for any reason, the light would be switched to red. At the same time, the cause for stoppage would be recorded on a log sheet on a clipboard placed by the machine. Also recorded would be the stop time and the start-back time.

These log sheets would be reviewed in weekly meetings to determine the most frequent causes of stoppage. Teams would be appointed to address these common causes and find permanent fixes.

Good Pinpoints

What makes for a good pinpoint—other than that it creates a clear focus? Good pinpoints should be:

- customer driven
- related to what you were hired to do
- worthy/valuable
- rally-able
- influenceable

Pinpoints must be customer driven. The selection is guided by understanding the customer's needs. If you are having trouble determining where to focus, just ask your customers what you should focus on. Then get to work on that.

Pinpoints should relate directly to what you were hired to do, or perhaps stated as why the organization or unit was created. These themes are often expressed in the company mission, vision, and strategy. Of course, that will quickly take you back to customer focus. This guideline will also steer you away from working on "training"—unless you are a training organization. It will keep you from working on productivity or efficiency. No organization was created to produce productivity or efficiency, but rather to produce a product or service productively and efficiently.

Pinpoints should be worthy and valuable. We don't want to choose a pinpoint that is really not worth our time and effort, such as one that will not even result in a return on investment. In fact, we are looking to make a significant impact (a paradigm shift) on the business. This is where the idea of the significant few versus the trivial many comes into play.

Pinpoints should be rally-able, meaning that we can be proud of the improvement when we accomplish it. Such achievements are the kind of thing you would tell

> **KEY POINT**
>
> I'm working in our Brazil office, conducting a 3-day Accelerated Continuous Improvement Workshop. Someone suggests "cross-training" as a pinpoint. I say that the Brazil office was not created to cross-train associates. "But," I asked, "Why do we need to cross-train our people?" The explanation was that we have some new equipment and systems that only one person knows how to operate. If that person is out for the day, we are not able to use the system or equipment to obtain the information we need to get back to the customer.
>
> Aha, "get back to the customer." Now we are getting somewhere! We do have an office in Brazil so that we can have faster responses to our customers there. So, response time to customers becomes the pinpoint, and cross-training becomes an action plan or project to improve response time.

an old friend about if you got together after not seeing each other for some time.

Pinpoints should be influenceable. This last criterion reminds us that we do not have to have "control" of the pinpoint. We only need influence. We really have control of few things in life. We think we have control, and we desperately desire control in our lives, but, for the most part, having control is only a dream. On the other side of the coin, there are few things in our life that we do not have influence over. We are very seldom just victims with no recourse.

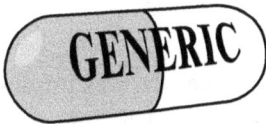

Generic Pinpoints

Here are several pinpoints that you can never go wrong with:

- **Timely**—(1) reducing the time it takes to "get one" (response time or lead time) and (2) being on time, as promised.
- **Quality**—How good is one when I get it?
- **Make business easy** for your customers. What are we doing that makes life more difficult for our customers?
- **Resource utilization**—Are we being good stewards of our materials and equipment?

Each of the above has unlimited potential for improvement!

The pinpoint must be precise enough to suggest a measure and to allow for goal setting, action planning, feedback, and reinforcement. These common terms are **not pinpoints**:

- **Cost reduction/cost cutting**—A focus here will result in an unending downward spiral (especially labor cost). Instead, focus on the causes of excessive cost: waste, downtime, and quality.
- **Productivity**—Productivity is not a rally-able pinpoint for the workforce. Productivity means working harder. Instead, articulate the focus as resource utilization, which equates to being good stewards of the resources available to us. Instead of making more per time, focus on how long it takes to make one (which is rally-able).

It should be pointed out that pinpoints are *not* chosen on an annual basis. Once identified, the work on the pinpoint continues until a high and steady level is reached. Many times, attention to a pinpoint lasts for years. (It is one of *the* performance areas important to the organization's future.) Some pinpoints emerge suddenly because of a problem or a change in the business, but most of the time, as the leader, you are able to see a new pinpoint that will emerge in the future—years before it becomes critical.

Leaders see farther, sooner, and more.

CASE FILE

Made-Right Cabs (MR. C)

My first visit to the cab company in Greeneville, Tennessee, was unsuccessful. I had been invited in by the HR manager whom I had worked with at another company. I was scheduled to give a 1-hour presentation on ACI to the leadership team. The objective was for them to determine if the approach would add value for them and if it resonated with their culture.

Upon arriving and getting set up in the conference room, I was told that the plant manager was running late. After 20 minutes or so, he arrived. He was obviously in a hurry, distracted, and making an attempt to squeeze me in before his next meeting. I suggested we reschedule at a time when we could be more relaxed, focused, and attentive. The group agreed and I left thinking, "I can really help these folks." (It is always my objective when working with an organization to help them concentrate their efforts and simplify their lives. This was another case where it seemed obvious to me that they were making it harder than it should be. It's not this hard.)

I tell this case story because after the next meeting with the plant manager Jay and his team, he got it. He really got it! In fact, he became an exemplar in leading an organization to implement the ACI methodology.

I loved the purpose statement of this factory that made cabs where operators of heavy equipment sit. Their purpose statement did not speak of making cabs, but of protecting people (drivers who sat in the cabs) and making them comfortable. They did this by making cabs that were rollover protective and comfortable.

In the 3-day workshop that followed, we went through the customary wrestling match to define the pinpoint, which became "made-right cabs." Made right meant that the entire process, from the paperwork that described the order to the loading of the cab on the truck for shipment, had no interruptions.

Made-right cabs:

- The work order was 100% correct, with no modifications required.
- All the components needed for the cab were readily available.
- There were no defective welds and no paint smears, runs, or scratches.
- The cab was ready for shipment as scheduled.
- A red tag accompanied the order and if any interruption occurred, the problem was noted on the card and that cab was designated as not made right.

The score for made-right cabs was calculated each day and posted on the plant scoreboard. Beginning from a level of around 50%, made-right cabs increased to 75% over the next 3 years. Word spread around the company, and leaders from other company locations came to see what was happening.

📝 **THE STORY**

What Is Leadership?

Jay, the plant manager and my friend, tells his experience this way in his writings called The Leadership Daily.

It was after I had known Russell Justice for a while that he shared with me the essence of leadership—the two keys. He said, "Great leadership is creating a focus and rallying your team around it." To that point, I had grown to respect and appreciate RJ, and he was recommended to me by colleagues I respected. But I have to admit I was disappointed with his answer—very disappointed. After all, everybody knows that leadership is about creating focus. Nothing new here. I shared my disappointment with him and that I already knew leadership was about focus; everybody knows that.

He said, "Sure you do, but have you ever really created focus? And I'm not talking about what is in your head. I'm talking about what your people say the focus is. If we asked 10 people around here, 'What's most important?' how many answers would we get?" I shared that we'd probably get 10 different answers. "Well," he said, "you haven't created a focus that your team is rallying around."

At that moment, I finally got it. Things started to change.

When you have completed pinpointing, you will have achieved a measure of like-mindedness, a term that I like. It represents a powerful concept and is one of those terms that sounds exactly like what it means.

"Then make my joy complete by being like-minded, having the same love, being one in spirit and of one mind."

—Philippians 2:2

KEY POINT

Often you begin with pinpointing, move on to the design of the next elements of your intervention, but find yourself circling back to the pinpoint. As you work on the next elements, you realize something is just not right about the pinpoint: It doesn't have the right focus or can be articulated better. At this point, you go back and spend some more time making sure the pinpoint is correct and everyone is like-minded about it. This cycling back can take place several times before you are rock solid about the pinpoint. It's critical that the pinpoint is exactly right because it is going to drive a lot of effort, energy, and time–and determine the future of the organization.

Get Focused–Stay Focused

I wish that getting focused was all there is to pinpointing: achieving focus and like-mindedness. However, there is a second part to pinpointing: staying focused. It's a real challenge to get focused and another real challenge to stay focused. It has been my experience that this is the second greatest cause of failure in organizations–right behind the failure to focus.

The "seduction of leadership" says that there are more and more attractions in business every day trying to seduce management away from their focus. That seduction occurs when the senior leader takes a trip and reads the business magazine in the back of the airplane seat. They come back home ready to implement the latest fad and throw the organization into chaos. Unfortunately, some organizations are managed by "seatback management." The battle line for our companies and our futures is drawn here.

What happens is that we get "off on the dotted line." (Doing anything that would not have to be done if the process was working correctly.) Consider Billy in Figure 1.1. His mother sends him to drop a letter in the street mailbox. On the way he gets sidetracked petting the dog, riding his bicycle, and visiting a friend. I sometimes ask team members to keep an "off on the dotted line" log for 4 weeks, in which they make a note each time they go off on the dotted line, why, and how long it took to get back on task. It's not unusual to find that 50% of the work day is spent off on the dotted line.

I once worked with Charlie, whose Maintenance Division of over 800 people had responsibility for thousands of pieces of equipment. They

Figure 1.1. We keep going off on the dotted line.

chose equipment reliability as their pinpoint and built the ACI plan to drive improvement in that area. However, as time passed, supervisors and mechanics in the division kept suggesting other areas for focus. As the leader, Charlie had to hold the line on equipment reliability. He would say to the team, "We have started climbing this mountain, and we intend to reach the summit. We are not going back down to the bottom of the mountain to start climbing up another trail." In other words, we're not going off on the dotted line.

Dr. Edwards Deming emphasizes this important principle in his "14 Points for Management," where his point 1 is "Create constancy of purpose." He also had "Lack of constancy of purpose" as the first of his "7 Deadly Diseases of Management" (Deming, 1986).

In the late 1980s while traveling with Dr. Deming as his host on a flight back to Washington, DC, I had the opportunity to ask him about why he made "Create constancy of purpose" point 1. I told him that it seemed to me that the first point must be a really important one. He answered, "If you're not going to create and maintain constancy of purpose, don't even start" (W. E. Deming, personal communication, 1980s).

Testing the Pinpoint–Putting It in the Spotlight

To make sure you have the right pinpoint, "put it under the spotlight." Ask the following:

- If we improved in this area, what would be the impact, the benefits? In other words, *why* would we do this?
- What would we have to do to bring about improvement in this area? In other words, *how* would we do this?

As you go through this testing exercise, you may find that you are not at the sweet spot. You may realize that one of the "hows" would solve 80% of the problem and tighten up your focus. In that case, you would make that how the new pinpoint and put it under the spotlight for testing. On the other hand, you may realize that even with significant improvement in the chosen pinpoint, you still would not be satisfied. In this case, you may have to choose one of the whys as a better or higher-level pinpoint. You can create an influence diagram of the components and move forward or backward until you are confident about that sweet spot. See Figure 1.2 for an example diagram.

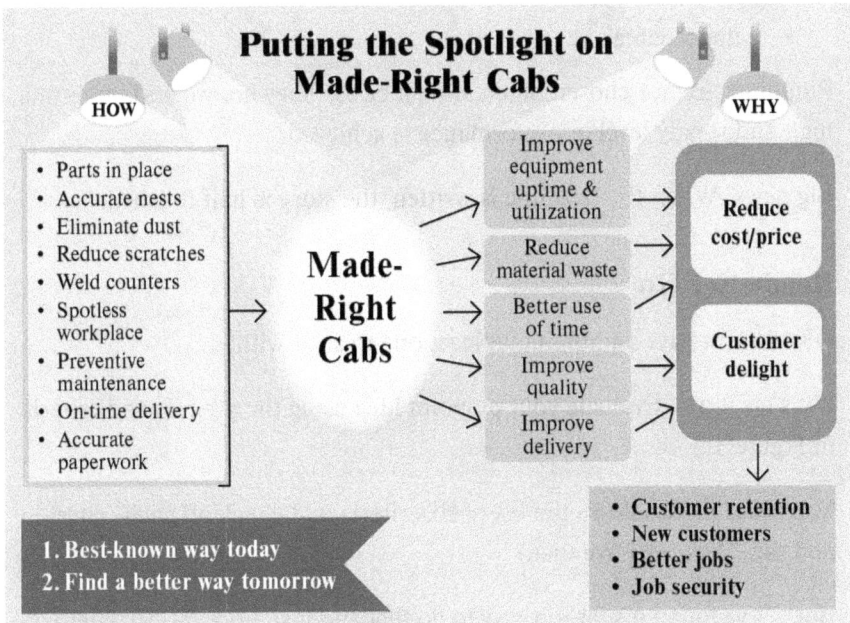

Figure 1.2. *An influence diagram with why and how helps validate the pinpoint.*

Focus/Pinpoint Summary

- Two Key Points—(1) **Get** focused; (2) **stay** focused.

- The single greatest cause of organization "failure" is lack of focus.

Five generic pinpoints:

- Lead Time: How long does it take to get one?

- On Time: When I get one, is it on time?

- Quality: When I get one, is it a good one?

- Make Business Easy

- Resource Utilization

Pinpoints should be:

- customer driven

- related to what you were hired to do

- worthy/valuable

- rally-able

- influenceable

Pinpoints are not chosen on an annual basis. They are worked on until a high and steady level of performance is achieved.

Big news: When the headline is written, the story is half finished.

Ready-Set-Go

OK, do you have a pinpoint, at least one to start with?

Let's try it out. Evaluate your pinpoint idea using the process and criteria in Figure 1.3.

You *could* just go down the list of six criteria and check off each criterion and say, "Yes, we have that."

But ... I've found it's far too easy to do that and just move on. So, after you check off the criteria, use the pinpointing considerations in Figure 1.4 to test the quality of your pinpoint.

Selecting the Focus/Pinpoint

Customer needs
_____
_____
_____
_____

Mission & vision
_____
_____
_____

Current priorities/ emphasis
_____
_____
_____
_____

1. Define criteria for pinpoint

2. List possible pinpoints

3. Develop influence diagram to illustrate relationship

4. Select pinpoint

5. Verify pinpoint against criteria

6. Put pinpoint under the spotlight

7. Make assignment to collect data

Pinpoint
_____
_____
_____
_____
_____
_____

Plan to collect data
_____
_____
_____
_____
_____

Criteria

☐ Is it customer focused?

☐ Is it what we were hired to do?

☐ Is it valuable, worthy? Will it make an impact?

☐ Is it under our sphere of influence/interest?

☐ Is it rally-able/braggable?

☐ Does it maintain continuity with current emphasis?

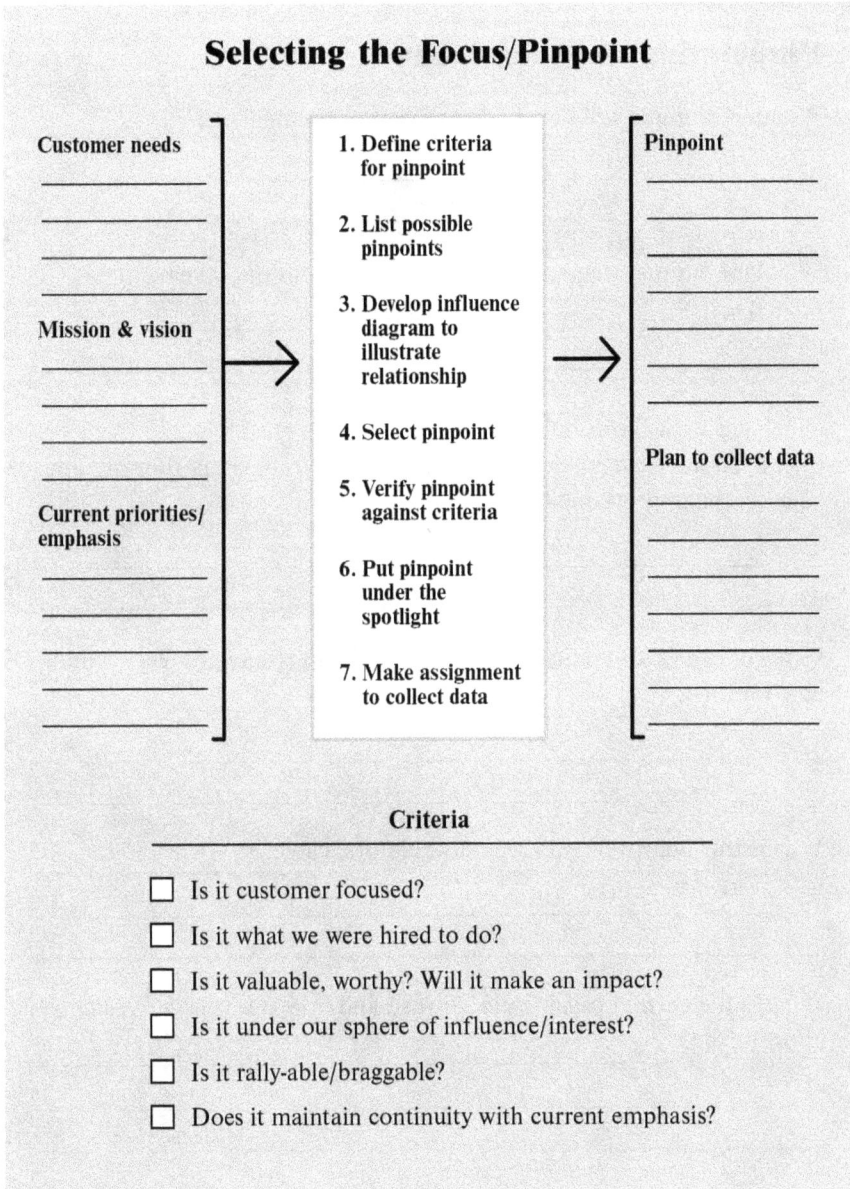

Figure 1.3. Use this process to guide you and move you forward with pinpointing.

Pinpointing Considerations

Which customers will benefit from this improvement? How? _____

_____

_____

Which of our major areas of accountability is this improvement
related to? _____

_____

What impact will this improvement have on the customer and the
organization? How much impact is expected? What is the "bottom line"
value of the improvement? _____

_____

_____

Can we make a difference in this process? Who else might feel account-
able for this? _____

_____

_____

Why is this improvement area rally-able/braggable? _____

_____

_____

What is the current organization thrust, and how is this improvement
effort linked? _____

_____

_____

*Figure 1.4. Evaluating the pinpoint against specific criteria tightens it up
and avoids oversimplification.*

See Appendix A for more examples of pinpoints and how they are used.

CHAPTER 2

Conduct a Kickoff to Engage the Workforce

Having achieved like-mindedness about what to focus our attention on, leadership uses a kickoff to share the focus and the challenge with the workforce and enlist everyone's help.

While a straightforward, businesslike presentation will do the job, you may want to add some drama to the kickoff to grab attention. You're sending the message that things are changing around here. It's going to be necessary to get outside the routine, and we want to have some fun while we are at it. For that reason, the kickoff might involve a video, a skit, an activity, or a rap. It might be held in a location not normally used for meetings—like on the factory floor, a stand-up meeting in the office, or at a movie theater.

As word gets out that "management is up to something with another one of those initiatives," everyone will want to know: What is being done? What is my part? What support can I expect? Who's the leader? When will we start? How will I know how I'm doing?

The kickoff is one of those times when leadership must get it right. This one is too important to blow it. You must obtain the necessary involvement and ideas of all associates. You have defined the challenge; now go get help. For this reason, there should be a dress rehearsal where each member of the leadership team participates.

It's important to recognize that the kickoff is the primary way to communicate the new initiative, but certainly not the only way. Following the kickoff, the messages about the effort must continue. These messages must be:

- completely consistent from every member of leadership (thus, the need to practice the message together)

- broadcast through multiple channels (every channel possible) to communicate the what, why, how, and progress of this initiative

- a steady stream of information and feedback from beginning to end of the effort

The Four Steps of a Kickoff

It's essential that the kickoff is conducted using the four kickoff steps as a guideline. Simplicity and clarity are musts. The four steps answer these four questions:

1. **What are we improving?** Explain why this focus area was chosen and why now is the time for action.

2. **Why is this important?** What are the benefits to our customers, to the organization, and to the workforce?

3. **What are we (leadership) going to do?** We are not asking you to do something we are not going to do ourselves.

4. **What are we asking you (workforce) to do?** We need your help. Otherwise, we will not be successful.

CASE FILE

Climbing the Pyramid of Relationships

To launch an initiative in Latin America to improve customer satisfaction and increase sales, we conducted a kickoff and workshop in Cancún with personnel from all the Latin American offices. The effort centered around the theme of climbing the pyramid of customer relationships. The workshop included a day trip to the pyramids of Tulum. The intent of the trip to the pyramids was to create a visual (and emotional) experience to plant and keep the initiative in the forefront of everyone's mind.

We returned from the workshop with a plan to keep a model pyramid (see Figure 2.1) on the conference room table in each office. The pyramid levels were labeled as Getting to Know the Customer, Establishing the Relationship, Growing the Relationship, Solidifying the Relationship, and Pinnacle. Each office was to identify key customers and place their logo flags on the pyramid at the appropriate level representing the relationship. At weekly meetings, the status of each customer would be noted, and the flag shown at the proper level. Successful moves to a new level were discussed and celebrated.

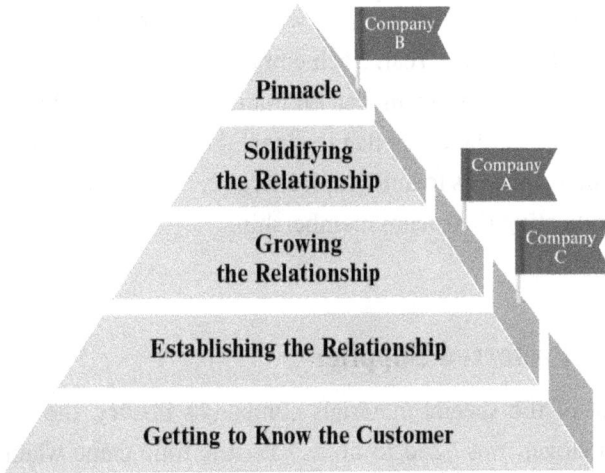

Figure 2.1. There are five levels of customer relationships.

📑 CASE FILE

Greeneville 200

Since January, 150 employees had been printing raw materials for the company's line of gift wrap. Then in April, another 150 employees, whose jobs involved converting the printed material, were called back to work to convert the printed material into marketable products. With 200 days of production until the November 1 deadline for the Christmas season, the improvement initiative was dubbed the "Greeneville 200" and given a NASCAR® theme.

The kickoff rally was held as the 1st shift was ending and the 2nd shift beginning. The crowd was warmed up by a stirring rendition of the "Star-Spangled Banner" and music by an all-employee band. The rock-abilly classic "Hot Rod Lincoln" was reworked into "Hot Rod Printin'," with the introduction line changed to say, "Mama said, 'Son, we're gonna be hurtin' if we don't step up that printin' and convertin'.'"

The plant manager, Scott, looking the part of a race car driver, was driven to the indoor stage in the NASCAR Busch Series pace car. He reminded the team that winning in NASCAR, just like in printing and converting, requires skill, speed, innovation, safety, and teamwork. He explained that the measure for the race would be feet of wrapping paper produced per dollar spent (ft/$) and that there would be three cars in the race: Last Year, Budget, and This Year. Each employee was given a Greeneville 200 Ticket and asked to write an idea for improvement on the back and turn it in. Before waving the green flag to signal the beginning of the race, Scott invited the employees to autograph a large car banner (with the company name) indicating their team membership.

CASE FILE

World's Preferred Supplier

For most of the special materials company's history, the customer satisfaction slogan was "2nd to none." Yet the time came when it became clear that slogan would not cut it anymore. Our push to turn that around was a preferred supplier initiative. This initiative was kicked off around the globe with the slogan "get on board" and a train logo.

To challenge teams worldwide to link in to the initiative, the company CEO filmed the invitation kickoff video from a train station in Paris. As the train pulled into the station, he stepped out onto the platform and shared his excitement about the initiative and its importance, saying that it was a hands-on, roll-up-your-sleeves effort that involved every employee. Then he invited everyone to "get on board."

Another video segment taken inside the bullet train in Japan showed the marketing president inviting everyone to take a seat, saying, "Form your team, identify a project, and get on board."

CASE FILE

Material Effectiveness

A plastics manufacturer chose the yield of PET plastic as the pinpoint. Because of the continuous operation process (24/7/365), it was necessary to conduct the kickoff with four separate crews (each consisting of more than 100 associates).

After the leadership team had practiced the kickoff (dress rehearsal), each crew was brought in early for the kickoff meetings. The leadership team presented the improvement initiative using the four kickoff steps: (1) What are we improving? (2) Why is this important? (3) What are we (leadership) going to do? and (4) What are we asking you (workforce) to do?

A couple of examples of "What you can do" were presented to prime the pump, and then the question was asked, "What else could be done to drive yield up?" The floor was opened for input. Microphones were available around the room for those sharing an idea, and the ideas were written on large posters for all to see.

Info cards were handed out, summarizing the four kickoff points and with space on the back to write ideas for yield improvement. Participants were encouraged to take the cards and give thought to what their team could do to help drive yield up. There was a box where the cards could be turned in while leaving and more cards were available for pick up.

Participants were informed about the idea of linking in (choosing a measure/project for their team that would drive yield) and invited to notify the leadership team when they had an idea to propose. (Note: See Chapter 3 for more on linking in.)

⬚ CASE FILE

Papermill Rap

To launch the improvement initiative at this Southern U.S. papermill company, four shift leaders who had participated in the design of the improvement plan put together a rap song. Each foreperson did a verse as the four points of the kickoff were covered. It was wildly enjoyed by the workforce, readily accepted, and not to be forgotten.

⬚ CASE FILE

We Deliver (New Stamps to the Post Office)

To start the project for delivering a new type of stamp to the United States Postal Service, a kickoff was planned where completing and delivering the order was compared to delivering the mail. The project emphasized the Persian Empire quote often ascribed to the post office, "Neither snow nor rain nor heat nor gloom of night stays these couriers from the swift completion of their appointed rounds."

As the factory employees gathered for the kickoff, the plant manager arrived dressed as a mail carrier on a mail-delivery scooter. The scoreboard was unveiled, showing a route from the factory to the post office, and the challenge given to anticipate, identify, and prevent problems along the path (e.g., dogs) that might result in missing one of the three goals of on-time, top-quality, and within-budget product.

CASE FILE

Make International Business Easy (MIBE)

To launch this initiative around the globe, 3-day workshops were conducted in offices in over 15 countries. The kickoff explained the need, enlisted the help of the personnel in the office, and began the process of identifying a pinpoint and developing an improvement plan.

CASE FILE

Knocking Down the Giants

In the process of beginning a new church, the leadership and congregation were faced with some giant challenges: purchasing 26 acres of land; designing the campus; and building, paving, and furnishing in phases to avoid debt. To illustrate the challenges, at a rally on the newly purchased property, large wooden giants were constructed and then pushed over by a bulldozer symbolizing the victory.

Kickoff Summary

- The kickoff is the mechanism to involve the entire workforce in a major improvement opportunity.
- The kickoff should be conducted in a way that sends a message that this is not business as usual.
- Opportunity should be given during the kickoff for the workforce to share ideas.
- Follow the steps for an effective kickoff by answering four questions:

- ○ What are we improving?

- ○ Why is this important?

- ○ What will the leadership team do?

- ○ What is the workforce being asked to do?

- Following the kickoff, a steady stream of communications must continue to emphasize the four questions and the appeal for everyone to get involved.

- These communications must be consistent over time and from every member of the leadership team.

Ready-Set-Go—Kickoff Checklist

☐ Build a compelling case for the need to focus on improving in this area.

☐ Develop a presentation following the four guidelines for a kickoff.

☐ Different members of the leadership team cover the four steps. This demonstrates the unity of the team. The kickoff should not be conducted by one person.

☐ Have a dry run/dress rehearsal of the kickoff. Consider recording it and reviewing it together for improvements.

☐ Bring all employees together for the kickoff (by shift if necessary; by video if necessary). Off-site is an option.

☐ Leadership conducts the kickoff.

☐ Talk about how performance might be measured. Say that there will be a scoreboard to make performance visible.

☐ Ask the question, "What are some of the things you and your team can do to help in this area?" As people give ideas, post them on a flip chart or screen for all to see. You want people leaving the kickoff with some "seed" ideas.

☐ Ask each supervisor/team leader to meet with their team following the kickoff and define the way their team will help—their link-in project.

☐ Talk about planned "link-in meetings" where teams will come and share with leadership what they have selected to work on, why that project was chosen, and the approach they plan to take.

☐ It's OK and effective to tell workers to be skeptical that this is just another management program that will go away soon enough ... until they see the evidence that it is real and here to stay.

☐ You are playing "Catchball" with the workforce (more on this in Chapter 3). Once you "throw the ball out," you have to listen to the reaction, to what people say they will do differently, and to determine if the right message made it through.

☐ The kickoff is a significant event and should grab attention, but make sure it is not bigger than the celebrations that will take place when major milestones are later achieved.

Help Teams
Translate & Link In

Focus & Alignment

Right along with getting focused, one of leadership's key jobs is the task of creating alignment with that focus. Focus and alignment are two of the vital few leadership elements that create the most leverage for the organization. In combination, they are powerful tools for creating competitive advantage and organization excellence. They also fulfill the basics of the Japanese approach called Hoshin Kanri, which has proven to have few, if any, competitors for achieving results. The goal is to be laser focused and purposely aligned and linked.

Associates are not the most valuable asset of
the organization; they are *the organization.*

Speaking the Local Language

With a clear pinpoint established and communicated to the workforce through the kickoff to all associates, the next challenge is to help teams translate the focus into their "local language" and to link in to that focus. I have found this to be an enjoyable and satisfying task, as individuals and teams begin to see how their work fits into the organization's focus, and the importance and necessity of their contribution.

Often the translation will be to dollars for senior management and to percentages (like percent uptime, percent yield, percent on time) for middle management. For frontline workers, it's about counting things, like the number of late shipments, errors in paperwork, the number of students earning an A, how many minutes it takes to complete a task, the number of employees who are fully qualified, or the number of patients returning for additional care.

Making Music—Not Noise

Helping teams to link in is the necessary step of creating alignment. Creating this alignment is like the difference between an orchestra still warming up (noise) and playing together (music). The difference is not the musicians (they may be world-class); the difference is the conductor—the leadership. Once the conductor stands up and announces the piece of music to be played, then everyone can play their own instrument. In this sense, we can say that it is leadership's job to orchestrate improvement. In a similar way, it is leadership's job to create the magnetic field flux for improvement, getting all the "iron filings" on the piece of paper to align us in the same direction (see Figure 3.1).

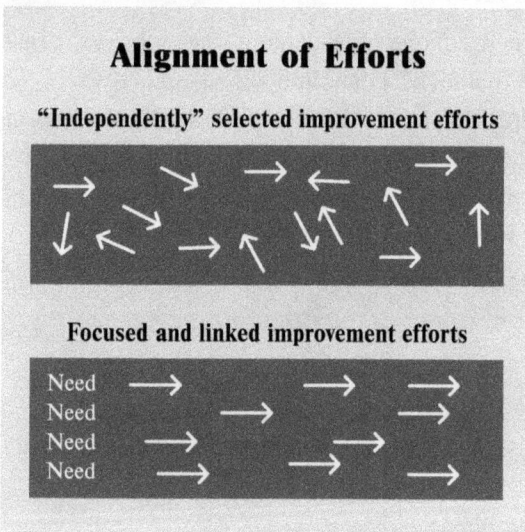

Alignment of Efforts

"Independently" selected improvement efforts

Focused and linked improvement efforts

Need
Need
Need
Need

Figure 3.1. Alignment eliminates conflicting efforts and creates a driving force.

How It Works

In the kickoff, you ask for the help of all associates, reminding them that success is not possible without their discretionary effort. Following best practices, identifying root causes of disruptions, finding ways to prevent disruptions, and generating innovative ideas all serve as ways to help drive the needed improvement.

The expectation is that each work unit will establish their own pinpoint—that one performance area that most directly impacts the organization's focus and overall pinpoint. At each level in the organization, the focus is sharpened to exactly how that unit influences the overall pinpoint.

The team may be able to identify this immediately, or they may need some help from leadership or a continuous improvement resources person. The leadership team may appoint some teams to take the lead in linking in as required. Do not force teams to link in, but encourage them to see what other teams are doing to get ideas. Offer help as needed to remove barriers and obstacles and to provide resources.

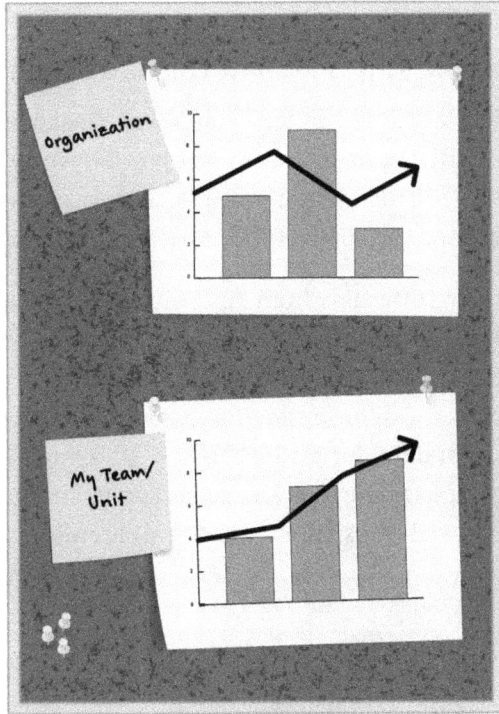

Figure 3.2. The best scoreboards show both the big picture (organization) and local picture (team/unit).

I like to call the overall organization pinpoint "The Big Picture" and supporting teams' pinpoints "The Local Picture." As the improvement system continues to develop, there should be a scoreboard like the one in Figure 3.2 that shows the overall pinpoint (organization) and the work unit's pinpoint (my team/unit).

✎ **TIPS**

Link-ins should be voluntary. You don't have to worry about making them mandatory as many leadership teams are prone to do. All it takes is a few link-ins to get the ball rolling. When these first few are recognized, shown attention, and celebrated by leadership, other link-ins will soon follow. The leadership team at an aerospace company had this concern, but 14 months into the program they had over 115 volunteer linked-in teams.

📑 CASE FILE

Make International Business Easy (MIBE)

For a global project aiming to "make international business easy" (MIBE), a link-in board was created where for each supporting project started around the globe, a satellite was added to the board. Each satellite showed the name of the project and a photo of the team.

Over 100 teams from 20 countries linked in to MIBE (see Figure 3.3). With a thrust to take away all of the complications and hassles that made life difficult for their customers, the teams' own pinpoints included response time for customer service or a product sample, adding freight forwarders to the shipping computer network, reducing response time to price requests, reducing order errors, same-day invoicing, and translation of technical literature into more languages—just to name a few.

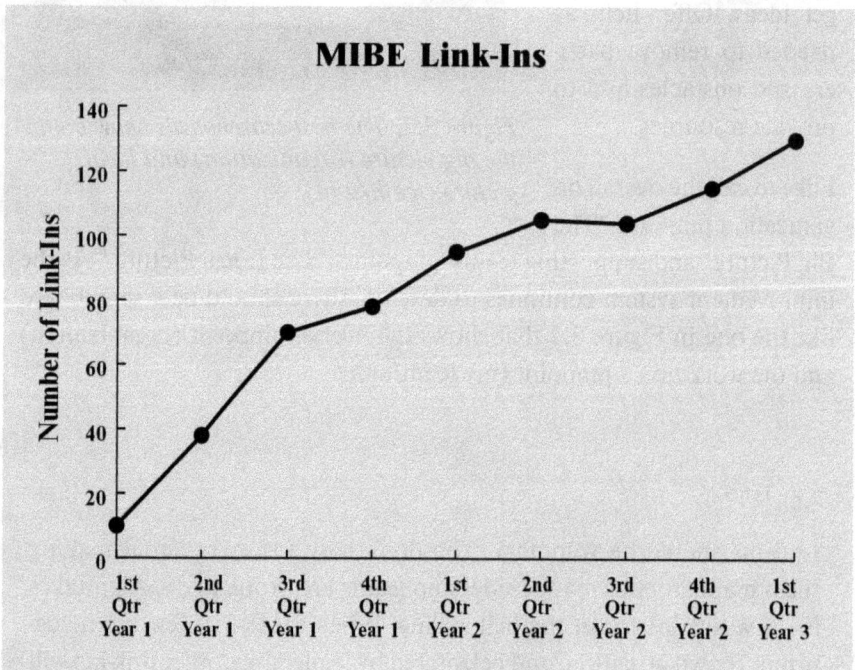

MIBE Link-Ins

Figure 3.3. MIBE link-ins rose steadily after the kickoff.

CASE FILE

Sales Revenue Growth

Link-in boards can also show the relationship between the various pinpoints. (Such a board is sometimes called an interrelationship diagraph.) For example, if the organization pinpoint is sales revenue growth, the local pinpoints chosen by the work units in the organization could be displayed as seen in the diagrams in Figure 3.4. Here you have a picture (worth a thousand words) showing how the various units are working together.

The diagram on the left shows the "themes" for each unit, and the diagram on the right shows the "language" spoken at the different levels as indicated by the way they measure improvement. As new units link in, they can be added to the diagram as illustrated here for teams linking in to the Maintenance pinpoint of equipment reliability.

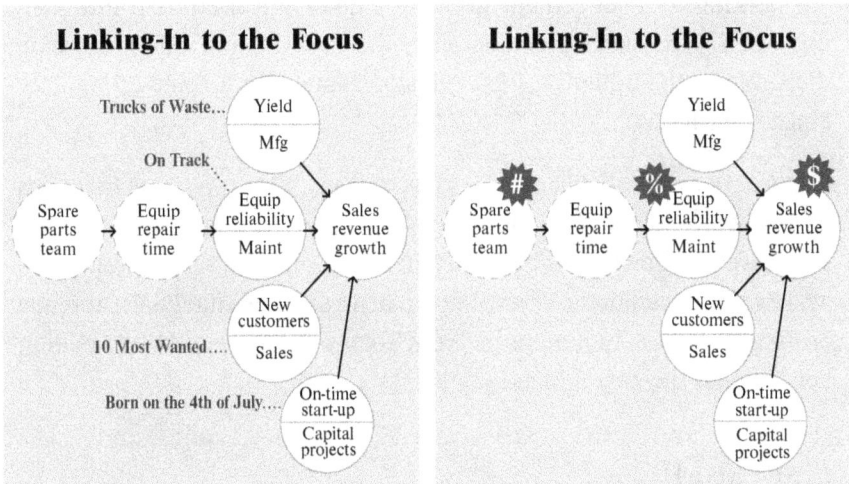

Figure 3.4. Themes strengthen link-ins and facilitate feedback and reinforcement. Link-ins should be expressed in the local language.

CASE FILE

Growing Sales in the Spare Parts Storeroom

Let's take a closer look at Figure 3.4 to examine the sales revenue growth link-in diagram and one of my experiences. Note the team called Spare

Parts on the left side of both panels of the figure. I was told by the continuous improvement consultant that this team had an incredibly good story to tell. So, I made a trip down to the storeroom where this team was located. It was not the most pleasant place to work as it was somewhat dark, stale, and in a basement with no windows. The men there said they remembered me from earlier seminars.

I asked what they were working on, and they said, "sales revenue growth." I pushed back on that a bit and said, "You're working in a storeroom. When did you last talk with a customer?" It was quickly obvious that the men there were getting agitated with me. One of them with an ominous look on his face took me over to a diagram similar to the one above for sales revenue growth. He said, "Justice, you just don't get it. We keep the spare parts to fix the machines. If we don't have the spare parts, they can't fix the machines. If they can't fix the machines and keep them running, we don't have product to sell, and there is no sales revenue growth. We are working on sales revenue growth!" I quickly realized that this team was "on it." They understood perfectly the objective and were focused on their role in the initiative. I left the storeroom with a smile on my face and a story to tell.

Can you imagine what happened in a company where you have hundreds of teams, over 15,000 people, all purposefully, consciously, working on sales revenue growth as it translated in their work area—all engaged to do whatever they can in their work area to drive sales revenue? Sales revenue broke a record for 13 months in a row. Total sales increased by $3 million per day over the next 3 years.

Catchball

The linking-in process is a form of the Total Quality Management (TQM) tool called Catchball, where two people play "pitch," throwing the ball back and forth to each other. This game of Catchball operates as follows:

- Leadership throws out the idea, the organization pinpoint that has been chosen. (This was done in the four-step kickoff described earlier.)

- Then the question is asked, "If this is the focus necessary for our future success, what would you do in your unit or team to help?"
- Next, leadership listens (catches the return throw) to see how the units and teams respond.
- If what they hear is not what they expected, if it does not line up with directly supporting the organization pinpoint, then the message must not have been either clear or compelling.
- At this point the "What are we doing?" and the "Why are we doing it?" must be articulated better. This is the work of leadership. This is what leaders do.

This process of kicking off the improvement (throwing the ball) and then listening for link-in responses (catching the returned ball) continues until there is tight alignment.

It also continues over time as more and more units and teams link in.

Note: Catchball is part of the Hoshin Kanri method. It allows you to align your company's pinpoint(s) with the actions of the people on all hierarchical levels of the organization.

Buy-In

You often hear leaders talking about "buy-in." The need for buy-in. We have to get buy-in. For me, the term buy-in can be one of those nonsense terms—it has no real meaning. It is just a vague notion or at best an assent from others that some idea is "OK."

I would define buy-in as giving freely of my time, energy, ideas, and efforts to an improvement initiative. It involves specific actions that are aligned with the objective of the initiative.

I submit that the linking-in process creates real buy-in. Buy-in is a tangible expression and visual representation of what a team or unit is going to *do* in support of the initiative. It's not enough to just say, "I like the idea." It's not enough for the leader to ask, "Are you bought-in?" More relevant is the question, "What are you going to do to make this happen?" (Note: There is an effective process for obtaining real buy-in. See Next Steps—Available From the Author at the end of the book.)

Leadership's Obligation

Anytime you ask someone to do something, you have two obligations: (1) Observe when they do it, and (2) acknowledge their effort and recognize it by saying "thank you." As Janis Allen's book title states, *I Saw What You Did, and I Know Who You Are* (Allen, 1990).

📑 CASE FILE

Golf Ball Wars

A golf ball producer was faced with two new competitors entering the market. The leadership team produced a movie (*Golf Ball Wars*) to explain the challenge of maintaining market share. After being challenged by the movie to find ways to maintain and retain customers even while new competitors were entering the market, departments chose the following link-in areas to focus on.

Table 3.1. Golf Ball Wars Link-In Areas by Department

Department	Focus
Packing	Bulk-pack miscounts
Molding	Changeover time
Core Components	Lost production due to lack of cores
Logistics	On-time customer orders
Stamping	Color consistency
Painting	Dirt-related defects

📑 CASE FILE

Material Effectiveness

At one of the world's largest PET plastics production facilities, the pinpoint was a material effectiveness initiative. Linking in began at the kickoff meeting, when employees were asked to share some ideas of what could be done in their area to improve material utilization. At the kickoff, employees were also provided a card where they could jot down ideas for improving material usage and turn them in.

After the kickoff, at weekly leadership team meetings, new link-ins were acknowledged and recognized. These link-ins included improvements from each Production Department (good practices and innovations), Material Handling (damage reduction), Maintenance (uptime), Development (process improvements), and HR (hiring and training).

Accounting took on the pinpoint for getting the needed yield information for feedback much faster. Before the Accelerated Continuous Improvement (ACI) initiative, yield information typically took about a week to collect and report. When the leadership team said they would like to have the yield information by noon on Monday so that it could be posted throughout the plant before the day shift went home, Accounting asked, "You mean Monday after the week ends on Sunday night?" Yes, the timeliness of feedback was critical. So, with considerable effort, Accounting did its part to help improve yield by finding a way to have last week's yield number by noon on Monday.

In the beginning, the Power Plant personnel felt they did not have a role in improving yield, saying that they provided the electricity and steam needed on a continuous basis. That opinion was held until the Monday leadership team review of yield showed a huge decrease on the chart. And the cause was power interruption. The Power Plant head came into the meeting saying, "I know. I know. I know our team's linking pinpoint—power reliability."

📝 THE STORY
Sleep Like a Baby

In the second year of practicing ACI, a chemical company focused on three major improvement opportunities (MIOs)—sales revenue growth, cost control, and customer satisfaction. All of these were aimed at being the world's preferred chemical company. To put legs to these three initiatives, the global leadership team (made up of corporate vice presidents for all administrative and support areas, manufacturing plant presidents and division heads, business and marketing executives and directors, and research heads)—about 25 people in total—were invited to attend a Linking-In Meeting.

From around the country and the globe, they came to the HQ building conference room. At the top of the wall in the conference room were posters for each of the three focus areas. Each poster illustrated how the pinpoint would be measured, along with other information about challenges and action plans.

Each company leader was asked to present their supporting pinpoint and how it would be measured on a graph. These supporting pinpoint graphs were added to the link-in diagram where each fit and connected by a ribbon to the appropriate major improvement opportunity. The wall turned into a masterpiece of how to turn ambition, intentions, and vision into a reality (Figure 3.5).

At the end of the meeting, the company CEO took it all in and reflected on what had happened there that day. He said, "I can't tell you how many times I have gone to corporate headquarters and presented our plans for the coming year—and then on the way back home wondered to myself if I had lost my mind, telling corporate that we would achieve those things.

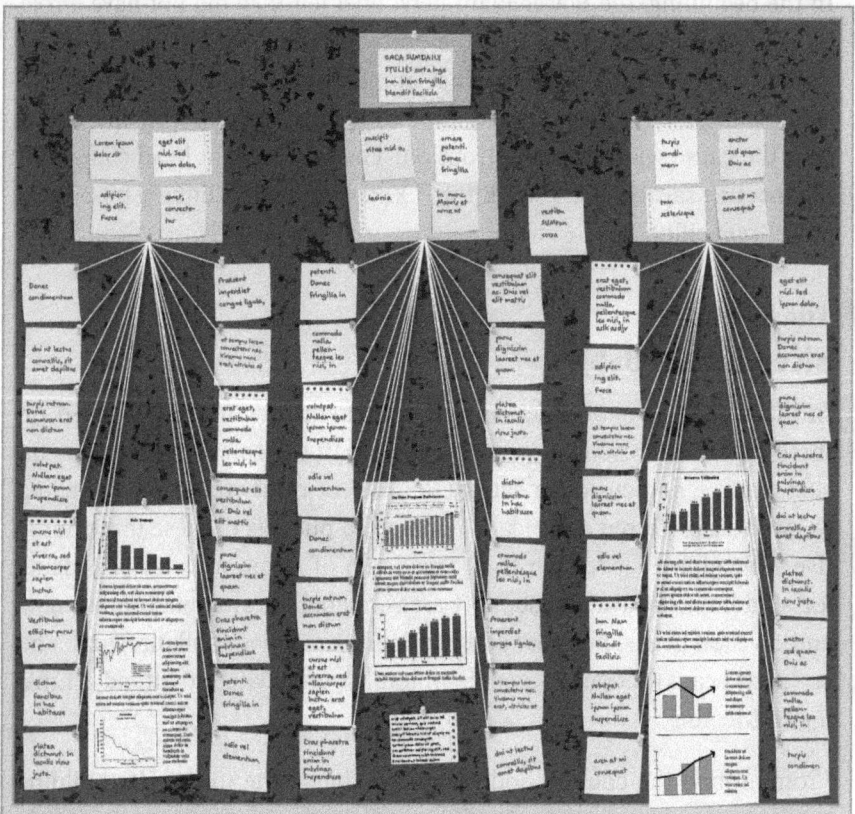

Figure 3.5. Wall-sized link-in diagram.

Not this time. Tonight, I'm going to sleep like a baby because I see right in front of me exactly how we will reach and, almost certainly, exceed our goals. Tonight, I'm going to sleep like a baby."

For myself that day, I will say it was a corporate meeting like none I had ever been to before. As it should be, with every pinpoint, every initiative, and every goal there should be a concrete plan to reach that goal. There should be a company guideline that no goals can be set without a detailed ACI plan of how to meet that goal. This day there were challenging goals and, at the same time, detailed plans from across the company of how they would be achieved. A process was in place—right before our very eyes—to take leadership beyond "wishing and hoping" to confidence that the plan would be met.

Translate & Link In Summary

- It's leadership's job, as the conductor, to orchestrate improvement.
- It's leadership's job to create the magnetic field flux that aligns all efforts.
- The goal is to be laser focused and purposely aligned and linked.
- Each work unit's link must be expressed in the local language.
- Although some teams may need to be appointed, voluntary link-ins are preferred.
- Offer/provide help to teams and work units in identifying their link-in.
- Each link-in should be recognized, celebrated, and shown visibly on a link-in board/diagram.
- Play catchball until the unit link-in is aligned with the overall pinpoint.
- "Buy-in" is achieved when a team/unit has shared what they will *do* to impact improvement.
- Remember the two obligations of leadership when asking for something to be done: (1) Watch for and observe action being taken; and (2) acknowledge the effort and say, "Thank you."

Ready-Set-Go—Teams Translate & Link In Checklist

☐ A specific time should be set each week/month when the management team will meet in a specified place to link in teams that have selected a project.

☐ Have a link-in board that shows the teams that have come forward with an idea/project that will help. Put a photo of the team on the board. Show the project and how it links. Celebrate the team's initiative in meeting and choosing a project. Have some refreshments.

☐ These link-in ceremonies may need to be virtual and sometimes even in the middle of the night for the leadership team, if the linking-in team is in another time zone.

☐ Let the team talk about their project: how it was chosen, some of the current problems, how progress will be tracked, and some initial improvement ideas. The team is working to vet their project here.

☐ Give the linked-in team a two-part scoreboard to put in their area. The scoreboard should have a space to display the overall measure for the organization pinpoint and a place for the measure of the progress of their linked-in project.

☐ Finally, and of utmost importance, make "linking in" a celebration event. Do not grill the team about details! If you have questions, jot them down for a later discussion.

CHAPTER 4

Develop & Carry Out a Management-Action Plan

Responsibility for the Plan

Of course, the primary action being taken by leadership is the development and implementation of the overall Accelerated Continuous Improvement (ACI) plan. It is this team that is launching the initiative.

Selecting the pinpoint and communicating it through a kickoff is entirely their task. They are responsible for the final design of the feedback system and the reinforcement system—both content and execution. The feedback system and reinforcement system will be discussed in detail in Chapters 6 and 7, with guidelines for leadership's participation.

As follow-through to the implementation, leadership is responsible for maintaining the momentum and replicating ideas that can be used in other parts of the organization. This can be done through the following:

- Hold a reunion where teams come together to evaluate the investment and return on their efforts and learn from each other.
- Host tours where teams visit each other's sites to observe best practices and learn.
- Organize a fair where teams share their stories through booths, storyboards, and presentations.
- Hold a symposium where performance excellence is showcased through the presentation of effective and innovative ideas.

Beyond the Plan

Beyond leadership's role in the design and implementation of the overall improvement plan, it is important for the workforce to see that leadership is investing in the improvement initiative, just like everyone else is being asked to do. Management must make an investment of their time, energy, and creativity toward the improvement by having their own team projects. By initiating policy changes and system changes, removing barriers, and tackling problems that can only be solved by management, leadership demonstrates its commitment to the initiative.

📑 CASE FILE

Exec-to-Exec

As part of the World's Preferred Supplier pinpoint, an Exec-to-Exec initiative with key customers was launched, where each company executive was assigned the responsibility to develop long-term personal relationships with a key customer.

- Customer participants were selected.
- The customer's interest in Exec-to-Exec was verified.
- An Exec-to-Exec checklist of duties was developed.
- Executives attended a training class on the initiative and their responsibilities.
- The Myers-Briggs Type Indicator was determined for each key customer executive participating in the initiative and information given to the assigned executive as an input to the relationship.
- Exec-to-Exec cards were developed that were similar to baseball trading cards. The front side showed the logo of the key customer, and the back side contained information about the executive: name, age, education, family members and ages, hobbies, and so on.

📑 CASE FILE

On-Time Program Performance (OTPP)

The leadership team of an aerospace company demonstrated their commitment to the OTPP initiative by taking these steps to achieve on-time program performance:

- Scheduled meetings every Friday to link new teams into the initiative, to hear about their project, and to encourage them.

- Created the **PAT**hfinders (Performance Acceleration Team) to help linked-in teams identify pinpoints and conduct projects.

- Calling themselves the Bureaucracy Busters, took a bite out of bureaucracy by eliminating time-consuming and unnecessary approval signatures: 20,000 annually by the time the project was finished. This initiative also reduced or eliminated a total of 150 reports and meetings.

- Dressed in safari shorts, hiking boots, and pith helmets during milestone celebrations in support of the theme: "On-Time Performance Will Get Us to Busch Gardens."

CASE FILE
Golf Ball Wars

A "Crackerjack Team" (CJT) composed of members of the leadership team, supervisors, and managers was appointed to oversee the development and implementation of the ACI plan.

In addition to working sessions, CJT meetings were held monthly to conduct link-in ceremonies. At the link-ins, the teams presented their plan to the CJT, explaining how their project linked to the overall company pinpoint, the value of reaching the goal, and some of the actions planned. This was a time for celebration and for encouragement by the CJT.

Coaches were established to help linked-in teams, provide resources as needed, and encourage the teams.

Management-Action Plan Summary

- The primary responsibility of the leadership team is to develop the ACI plan and to oversee its implementation.

- Beyond that, other actions may be needed and can only be taken by the leadership team or members of the team. Actions might include policy changes, major expenditures, and removal of barriers.

- The leadership team should have one or more projects to send the message, "We are not asking you to do anything that we are not doing ourselves."

Ready-Set-Go—Management-Action Plan Checklist

☐ Solicit input from the workforce about specific ways that the leadership team might support the initiative, such as ways to eliminate barriers and complications to getting work done by updating, changing, and eliminating policies and procedures that are not needed or helpful. Ask, listen, take notes, and follow through with actions.

☐ Schedule a meeting outside of regularly scheduled meetings with the intended outcome being a list of action ideas that the leadership team could do to enable and support the initiative and a selected action item as an initial project to be taken on by the team.

☐ Link in the project just as with other teams.

☐ Make the ideas list a living list. Make it visible for the team, and add to it as new ideas are generated or come to mind. Check off items as projects are completed.

☐ Schedule time each week to work on this and future projects, either in a regularly scheduled meeting or a special one.

CHAPTER 5

Teams Work to Improve Processes

Author's note: Since there are numerous excellent sources for methods and approaches to process improvements, I'm not going to address any specifics of process improvement here. I will share a few principles and some tips for the leaders' role in this component.

Eventually we must stop talking about what we are going to do and go do it. Yoda told us that there is no trying, there is only doing. And James tells us in the Bible to be doers and not only hearers.

"You can't build your reputation on what you are going to do."

–commonly attributed to Henry Ford

Ultimately, all improvement takes place one project at a time, by employees designing a new process or focusing on an existing process, identifying root causes, and preventing their reoccurrence. This work is the heartbeat of improvement.

Do It Right the First Time

Do you remember the old Japanese saying, "Do it right the first time"? Well, it's actually not an old Japanese saying. When I was leading a workshop in Tokyo once, I mentioned this old saying. I was quickly rebuffed by the local quality guru, who said that no self-respecting Japanese person would say such a thing—it was "nonsense." He pointed out that I had been talking about "continual improvement" all week and asked me which one I wanted—continual improvement or right the first time—pointing out that if you did it right the first time, there would be no need for continual improvement. Point taken. He went on to say that the Japanese way was to "Do it the best-known way today and find a better way to do it tomorrow."

Every Employee Has Two Jobs

This notion of "best-known way today and better tomorrow" leads to the principle that every employee has two jobs—daily work and improvement work. Do daily work with excellence by following good practices and being attentive to details. At the same time, while doing daily work, constantly think about and look for ways to improve the process through innovations or by reducing problems.

There should be time set aside every week for the team to work on improvements, but most of the scrutinizing and probing to identify problems and the generation of ideas and innovations comes in the workplace—at the machine, the desk; in the lab, the store aisle—while doing the work. In that sense, Job #2, finding a better way, is not something extra that has to be done, but part of the real job all along.

🔑 KEY POINT

The workforce embracing this notion of two jobs is a fundamental driver for accelerating improvement and achieving excellence.

Root Causes, Good Practices, and Innovations

Through the years, I've been amazed at the root causes identified, good practices developed, and innovations generated by a work team once they turn their attention to improving their processes and giving of their discretionary effort. Take a look at these examples from the case files.

CASE FILE

Got More Milk

Increase the amount of milk per cow per day through hoof trimming, soil testing, crop rotation, nutrient planning, supplements planning, time of day for feeding and milking, cleanliness of quarters, recovery time in the sick lot, particle size, magnets in rumen, music, and loafing barns.

CASE FILE

Press Uptime

The ways to prevent ink problems when starting up the printing press can be defined by taking the time to observe in detail every upset and by seeing the upset as a chance to find a way to make the process work better. Causes identified included feed adjustment, ink/water balance, roller settings, grippers, plate changes (scratch, file wrong), short runs, paper weight, moisture, flatness, welded edges, and static.

CASE FILE

Fill the House (Barter Theatre)

In 1933, with the United States in the middle of the Great Depression, the people of New York could not afford to pay for theater tickets. Theaters went dark and actors found themselves out of work. At the same time, farmers in Southwest Virginia were stuck with crops they could not sell. That's when Robert Porterfield from Southwest Virginia came up with his genius of an idea: Bring actors to Abingdon, Virginia, to barter their performances for the farm goods from the local farmers. The New York actors could enjoy performing and be fed from the admission proceeds. The farmers could make use of their vegetables and enjoy a good theater performance. Payment for admission was vegetables and fruits, hence the name *Barter* Theatre. Porterfield said, "With vegetables you cannot sell, you can buy a good laugh" (Barter Theatre, n.d.).

That now-famous theater operates today as a treasure of the region, but it still has challenges to face and solve. One of those challenges for a regional theater is to fill the house. A key to success is to keep the seats full throughout the season. That pinpoint—Fill the House—became the challenge. And the action ideas included the following:

- Free passes given at random to shoppers at regional malls. (To help support the theater, the malls give 1% of Barter Promotion Day sales to the theater.)

- Free passes given at random to fans entering nearby Bristol Motor Speedway (BMS). (In support of Barter, BMS gives 1% of the gate from sponsored races.)

- Dressing the "front of the house" with spiffy uniforms for staff.

- Constructing additional easy, convenient parking for guests.

- Training for all box-office personnel in consultative selling.

- Taking short plays and skits off-site to schools, business meetings, conferences, and local events like the Bristol Race weekend.

- Establishing a Frequent Patrons Program, with a "trading card" for each patron showing their information (name, photo if available, seating preference, refreshment choices). Cards to be given to all staff, who were then required to memorize them.

CASE FILE

Pizza Heat Bags

When spending over $40,000 per year on heat bags for pizza delivery, what can be done to reduce the cost?

- Redesign the bag for increased strength and tear resistance.

- Bag-handling training—have operators make a video of how to safely and carefully handle heat bags.

- Require delivery drivers to turn in a heat bag before picking up another order.

CASE FILE

10-Most-Wanted Customers

Having identified 10 customers who had never placed orders with the business, a plastics manufacturing company created a 10-Most-Wanted

Board. Each targeted customer had a section on the board for information collected on that customer, along with the customer logo.

One piece of intelligence on the board was a newspaper article about the daughter of the owner of one of the target customers. The daughter had performed as Clara in the town's holiday *Nutcracker* play.

Seeing this article and realizing that he would be attending a Kiwanis meeting with the daughter's father and the owner of the target customer, a sales rep struck up a conversation about the *Nutcracker* performance by asking "Clara's" father if that was his daughter.

The rest is history, as that conversation led to business discussions and a new (captured) customer for the company.

📑 CASE FILE

We Deliver (New Stamps to the Post Office)

Using the theme of "We Deliver" (like the post office), the company asked what deters the mail carrier from delivering? Since it is "Neither snow nor rain nor heat nor gloom of night," then the assertion was, it must be the dogs. With that in mind, cardboard-dog cutouts were made available, and employees were encouraged to identify potential problems and write them on the cardboard dogs. The dogs were then placed on a scoreboard along the path that illustrated the journey from the factory to the post office. Teams were formed to address each potential problem, and once resolved, the cardboard dog (problem) was moved to the "kennel."

Putting All the Elements Together

We have come far enough now to begin putting the elements together to illustrate the overall approach. Let's look at two examples that include all the elements we have discussed so far, along with the elements still ahead (slightly shaded). At this point, you will get your first look at the overall process.

📖 **CASE STUDY**

Improving Container Quality

Situation	• When damaged drums of dye arrive at the dock for shipment, it sets off a chain of "off on the dotted line" activities that are costly, frustrating, and detrimental to customer satisfaction. • Calls are made to inventory management to assign additional drums to replace the damaged ones. • If there are no available additional drums of the product, the order will ship short. • The customer may say that if the complete order can't be filled, they will go somewhere else. • The truck driver waiting for the load is upset because they are losing time and money waiting for replacement packages to be found, checked, and loaded. • The warehouse workers are working overtime to get the order out and showing up late to coach their Little League baseball team. Big problem!
Focus/ pinpoint	Container quality
Kickoff	• A meeting is held in the warehouse with all Warehouse workers plus representatives from Manufacturing, Production Planning, In-Plant Trucking, and Packaging Engineering. Also attending are a truck driver and a customer to share the frustrations they have experienced. • Historical results are shown, with a baseline of 93.5% shippable packages arriving at the docks.
Translate & link in	• Teams are initiated to work on causes of the identified problems—the things that would make the package unshippable. Warehouse workers volunteer for the teams. • The teams focus on product/dye on the outside of the drum, dents or scratches or punctures in the drum, label crooked or ink smeared, not sealed properly, incorrect weight.
Man-agement action	• Leadership conducts kickoff, designs scoreboard, plans and executes reinforcement. • Sets up 4-hour weekly meetings for problem-solving teams and appointed team leaders (Job #2). • Provides problem-solving training for leaders and team members. • Designates 15 minutes at the beginning of each shift for review of package quality–going over unshippable drums from the previous day and the reasons. Sharing newly identified good practices. • Initiates weekly meetings with Warehousing management, Production, Production Planning, Trucking, and Packaging Engineering. When needed, packaging vendors are included.

Improve process	New ideas: • New calibration system for scales. • New label and non-smear ink. • Air/fan system to blow dye dust away from packages. • Drum-handling training. • Team with vendor to redesign seals.
Measure & feedback	• Percent shippable on a weekly basis. • Measured at the shipping dock as packages are inspected using a checklist. • 4' x 8' plywood scoreboard at the docks. Includes weekly graph of percent shippable, baseline (93.5%), goal (98%), current year-to-date, and best week ever shown.
Reinforce & celebrate	• "I Was Helped By" chart next to the scoreboard where Warehouse workers could write the name of anyone who helped them do their job or helped in increasing the percent shippable, including what they had done. • Best-ever weeks and each 1% year-to-date improvement in shippable packages celebrated with food and refreshments at the end of the shift. • When year-to-date percent shippable was reached halfway to the goal of 98%, pocketknives engraved with "We cut out package damage!" were given to all Warehouse workers. • When the goal of 98% was reached, a celebration was held in the warehouse with some invited customers. Representatives from Production, Production Planning, Packaging Engineering, and Trucking also attended. Drum, Label, and Seal representatives were also invited. A customer shared how they had seen a difference in fulfilling their needs.
Results	• 98% shippable was reached by mid-year as a result of implementing the improvement ideas and the discretionary attention to detail given by all involved. • By the end of the year, the graph had "topped out," with the points on the graph all being 98% or more (see Figure 5.1). • For year 2, the scale on the graph was changed from 80% to 100%, to 98% to 100%. • In year 3, *percent* shippable no longer really made sense, so the scoreboard changed to the number of unshippable packages each week (some with zero) and the year-to-date number compared to previous years.

Note. Shaded rows discuss elements that we will cover later in the book.

Container Quality

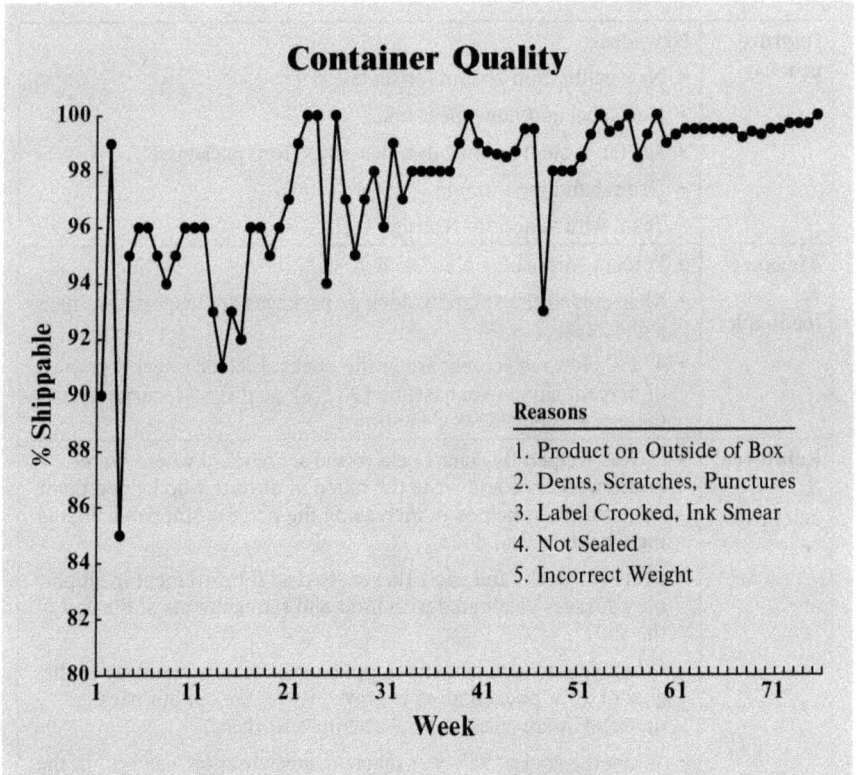

Reasons
1. Product on Outside of Box
2. Dents, Scratches, Punctures
3. Label Crooked, Ink Smear
4. Not Sealed
5. Incorrect Weight

Figure 5.1. Shippable containers improved to the point that the scale on the graph was no longer helpful.

CASE STUDY

On-Time Delivery of Supplies: Purchasing Department

Situation	Aside from raw materials needed for manufacturing, many other materials, supplies, and spare parts are required for daily production. These items are called code orders.
Focus/ pinpoint	On-Time Delivery of Code Orders
Kickoff	The Purchasing Supplies Group kicked off the effort with a 3-day workshop where the approach to be used for improvement was presented and the team developed an intervention plan.
Translate & link in	• The Purchasing Supplies Group was all "linked in" as a result of the kickoff workshop. • A presentation explaining the project was put together for supporting teams and individuals—vendors, Stores (warehouse), supply/material/spare parts users, Packaging Engineering, and Trucking.

Manage-ment action	Management instituted a weekly team meeting focused on pulling away from day-to-day activities (Job #1) and providing time for improvement (Job #2).
Improve process	• Worked with users (Materials, Supplies, and Spare Parts) to update all product specifications. • Developed a new form and process for updating specs as soon as a change is made. • Started a "clean order" report that showed what percentage of requests from users were completed correctly. "Dirty" or unclean orders were logged, tabulated, and feedback given to those submitting the orders. • Video put together to emphasize the importance of a "clean" order and how to prepare one. • A "double check" of automatic-release orders instituted to make sure specs were up-to-date before release. • Developed a weekly list of "anticipated late" orders that could be addressed before a problem arose.
Measure & feedback	• Number of overdue orders (weekly count). • Plotted on a bar graph on the scoreboard in the Purchasing Team meeting room.
Reinforce & celebrate	• Lots of praise in team meetings as specifications were being updated and problems being solved. • Thank-you notes sent to users for cleaner orders and to vendors for improved delivery. • Lunch celebrations as milestones met. • A "We Made the Elephant Dance" celebration (meaning we achieved something that seemed very difficult or impossible) was held, with cake and elephant buttons. • Less work for the Purchasing Team, Stores, vendors, and users as errors and expediting greatly reduced.
Results	• Baseline average number of overdue orders per week: 350 • Milestone goals met: 250, 200, 100, 50. • See Figure 5.2. • Increased vendors having appointments from 40% to 90%. • Annual purchasing savings: $800,000.

Note. Shaded rows discuss elements that we will cover later in the book.

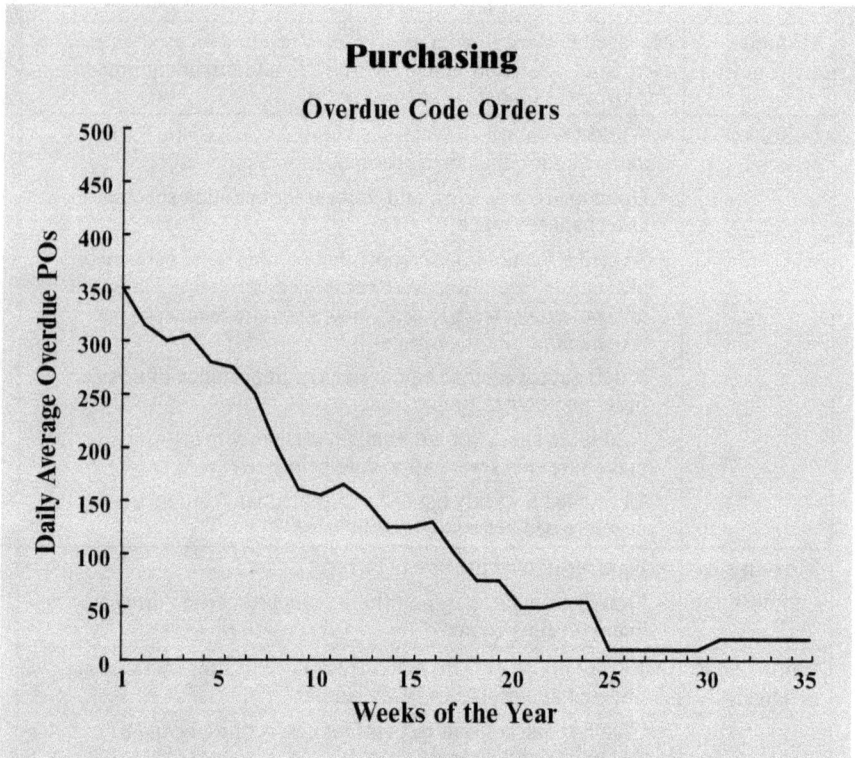

Figure 5.2. *Late purchasing supplies improved from over 300 to less than 50 during the year.*

Improve Processes Summary

- Every employee has two jobs:
    1. Do it the best-known way today.
    2. Find a better way to do it tomorrow.
- Process improvement takes place by:
    1. Executing good practices.
    2. Identifying root causes of problems and developing permanent fixes.
    3. Process innovations.
- The leader's role in process improvement is to ensure that a systematic process-improvement process is in place, that time is available on the job for process improvement, and that coaching resources are available to assist.

Ready-Set-Go—Teams Work to Improve Processes Checklist

☐ Time must be provided in the work schedule for working on improvements (Job #2).

☐ A systematic process-improvement approach should be selected by the organization.

☐ Coaching/consulting resources should be made available to teams and individuals serving as continuous improvement coordinators. Full-time is best, part-time if necessary.

☐ When a team has linked in, offer them training on the improvement process. Also offer coaching/consulting help.

☐ Don't give training until the team has invested the time to get together and identify a project and come forward to link it in. Teams earn the training.

Now What?

At this point, most people would be thrilled. They would be saying, "What else could we ask for? Yahoo! We have arrived! We have a clear focus for improvement. We have shared that focus with the workforce. Teams are determining what they can do to help drive improvement, and they are beginning to work on improvement projects. And we have a plan to show our involvement in the initiative."

Unfortunately, after you have agreed on the goal, have all the right people on the job, and have provided all the information and tools needed ... your work is not done. Two components are still needed to bring out the discretionary effort in the workforce.

How we *respond* to the improvement efforts has as great an effect on results as anything that has been said or done before the work began. In other words, what happens *after* people give their effort and improvement begins to happen?

How we respond to the improvement efforts has
as great an effect on results as anything that
has been said or done before the work began.

The job of a leader is not just to *tell* people what to do. Far from it.
Remember the two obligations? Anytime a leader asks the team or an
individual to do something, the leader has two obligations:

- Observe whether it is done.
- Recognize when it is done.

One executive who headed up an organization of over 10,000 people once
told me that his leadership style was to "tell people what to do and then
leave them alone." He was doing this under the guise of empowerment.
That sounds really good, but the only problem is, it doesn't work. If you
tell people what to do and leave them alone, soon enough their initiative
and effort will wane, as there is no acknowledgement or encouragement.
(The second law of thermodynamics, commonly called entropy, says that
without intervention, systems naturally progress to disorder or chaos.) In
other words, left alone, thing get worse.

This is where the challenging work begins. Most leaders spend their time
trying to motivate people before they do things. This is where your worldview of what it means to be a leader has to change. This is where we
are going to take leadership up to a new level by emphasizing these two
after elements.

This is where competitive advantage is created over other organizations
that don't follow through on what are the missing elements in most leadership systems. Those missing elements are *feedback* and *reinforcement*,
which are discussed in the next two chapters.

CHAPTER 6

Measure Progress & Provide Feedback

Let's Get Serious! If You Can't Measure It, You Can't Improve It

It's been said that if you can't measure it, you can't improve it. But that's not really all that bad news, because you can measure *anything*. When you talk about something, it often involves having measured it in some way.

For example, if I tell you that I like the shirt you are wearing today, that means that I have measured it in some way. I've examined the color, the pattern, the style, and "measured it" to be nice.

All Measures Are Wrong, and Some Are Useful

Another somewhat comforting fact is that "all measures are wrong, and some are useful." If you took out a ruler and measured a tabletop, the size you came up with would be an approximation. It is never possible to get the exact measure, even with the most sophisticated devices. Even the most advanced measuring devices have some degree of uncertainty or error. This is due to various factors, like the limitations of the instruments and the inherent variability in the objects being measured. In science and engineering, this is why we often talk about measurements in terms of their precision and accuracy, and why error margins are always included.

No Data, No Complaint (Bring the Data to Get the Platform to Speak)

I like to have this rule for team meetings: "No data, no complaint." Too often we hear in meetings complaints like, "The Maintenance Department never gets back to me." Such a statement begs for data. How often does this happen? When was the last time it happened? How many times did it happen last month? Have they ever gotten back to you? How many days did it take? What is the impact on your operation when they delay getting back to you? Collect the data, and then come back to the meeting and you can have the floor. Having the data gets you the platform to speak.

Measures Go Through (at Least) Three Phases

It has been my experience that measures go through three phases.

1. First you have a measure.
2. Over time, you get an accurate measure.
3. Eventually, you have a useful measure.

This means that you don't have to wait until you get the perfect measure to initiate the improvement. You start with the best measure you have. Plot and display it for all to see. If it is not correct, the workforce will tell you. Often, they will suggest ways to improve the measure.

I like the story of the three umpires in the bar after a game.

The first umpire says, "I call 'em like I see 'em." The second umpire says, "That's not right. I call 'em like they are." The third umpire speaks up and says, "Neither one of you knows what you're talking about. They are not anything until I call 'em."

Just call them and get started.

Measures Must Be Translated Into the Local Language

In Chapter 3, we discussed "translating" pinpoints (measures) based on the level within the organization. Dollars for senior management, percentages for middle management, and counts for the front lines.

While financial measures could be appropriate in some cases, they are usually not best for organization-wide pinpoints where the entire workforce is being rallied for improvement.

Baselines

When setting up measures, put them in the context of past performance or a baseline. Future measurements all find their place against the baseline—better, worse, or the same.

End of the Day

Associates should know how they did—their score—at the end of each day. If performance isn't visible to your teams until the next day, a week, or a month later ... or not at all, you have tremendous opportunities for improvement.

The Measure Finalizes the Pinpoint

It's important to note that establishing the measure finalizes the pinpoint. A worker may not really be sure what has been chosen to work on (the pinpoint) until they see how it will be measured. Then the focus and target of improvement becomes much clearer.

Four Types of Useful Measures

There are four types of useful measures, each of which is best at times. There are the objective measures of count and compute. And there are the subjective measures of rank and rate.

Count as Events Occur—A Good Practice

The best measures are observed and counted/tabulated as they occur, not computed after the fact or reported from the accounting system (often days or weeks later). Measures that we personally handle create more interest and are more rally-able.

For example, let's take hotel occupancy from Table 6.1. I suggest that a count of the number of empty rooms is more meaningful and rally-able than the percentage of rooms empty. You can't touch or see a percentage. You can go see and count an empty room. With the proper graph or scoreboard design that visually emphasizes rooms, you will be able to see a room being occupied.

Let's take another example from the list—a maintenance department and the number of machine stops per day. Accounting and engineering departments might suggest that it would be better to measure the number of minutes down for each machine and multiply that by the lost value for being down to get a dollar value for lost production per day. While that is a measure that should be tracked by leadership, it is more complicated and not my recommendation for the scoreboard. Mechanics do not fix minutes or value; they fix machines. The count is so simple, immediate, and visible: running or not. As the number of machine stops per day goes down, a transition to a more sophisticated measure (like number of minutes down) will likely take place.

Here are examples of events that could and should be counted (some of them as they occur).

- A project that is late being completed—not counted at the end of the month, but when the due date passes. Add it to the late-project list and launch an investigation to determine why.
- A student drops out of school—a person lost in the process, not a statistic.
- Empty seats at a theater show or performance.
- Machine break down—red light above the machine is turned on and stoppage logged.
- Call on a customer—score of the call recorded based on the sales-call scorecard.

Computing a Measure

Computing a measure means that a specific method or formula is used to determine the value. Examples of computed measures would include throughput, market share, defect density, employee turnover, cycle time, revenue per employee, and lead time for delivery.

Subjective Ranking

Subjective ranking involves placing items in order relative to other similar items based on some specified criterion. The criterion could be importance, value, or likeability. Examples include the value of supplier communication, the taste of french fries, the enjoyment of a theatrical play, or the best college for recruiting new employees.

Subjective Rating

Subjective rating is a type of evaluation based on personal opinions, feelings, and perceptions rather than objective measurements. Subjective rating is used when individual experiences and judgments are important. It is generally done on a scale of 1-5 or 1-10. Some examples are pain assessment in healthcare, taste of a food, or beauty of a sunset.

Table 6.1. Meaningful Measure Examples

Organization	What to measure	How to measure
Bauxite mining and refinery	Equipment utilization	Percent uptime
Chemical manufacturing	Cycle time	Hours per batch
Emergency room	Wait time	Minutes
Engineering projects	On-time completion	Number late (or behind)
Engineering school	Yield	Percent enrolled who graduate
Golf ball manufacturing	Market share	Percent share
Hotel	Occupancy	Number of empty rooms
Maintenance	Machine stops	Stops per day
Medical clinic	Quality of visit	Scorecard (18 points)
Pharmaceutical manufacturing	Throughput	Batches per week
Pizza store	Remakes	Number remade
Quick service restaurant	Time to serve	Seconds
Restaurant	Frequent diners	Number of visits per month
Shipping	On-time shipments	Daily leftovers
Theater	Attendance	Number of empty seats
Trucking	Cube usage	Percent of truck volume filled
Trucking	Damage	Number of items damaged
Warehouse	Package damage	Number of items not shippable
X-ray department	Repeats	Number of repeats

Measurement, the Ultimate Question—What's on the Y-Axis?

In establishing the measure for an improvement thrust, the ultimate question is, "What's on the y-axis?" How will we know if we are winning or losing? The *one* measure that will be used to determine success: the bottom line. The title of the graph (the pinpoint) is important because it tells us the subject of improvement, but the measure on the y-axis nails down the objective and challenge (see Figure 6.1).

Figure 6.1. *"What's on the y-axis?" is a question that nails down the issue.*

One of the leaders I worked with over the years often told me that the most important thing I ever taught him was to ask the question, "What's on the y-axis?" Some of my valuable contributions as a consultant to executive teams have been to ask this question and as a result, *stop* initiatives. The hundreds and thousands of members of these organizations have no idea how I made their jobs and lives easier by stopping initiatives with this question.

For example, I observed through the years that senior leaders like to do two things: (1) reorganize and (2) train. It is not uncommon for a vice president to come in with a reorganization idea, such as creating a new position and moving others around. When this happened in a leadership team meeting, I would go up to the flip chart, draw the x- and y-axis of a graph, and tell them that I have just three questions:

1. There must be something that is not performing at the desired level, and we want to see it improved by reorganizing (or with this training program). What do we want to see improved? Let's put it above the graph as the subject of improvement.

2. How do we measure it; how will we know if we are improving? Let's put it right here on the y-axis.

3. Who wants to sponsor this project? Let's write your name on the bottom of our chart as the sponsor.

More than once when I did this, the person suggesting the reorganization or training program would say, "I hate it when you do that, Justice. Just forget it!"

One Measure

A tendency by leaders, engineers, and accounting is to resist having just one measure, and for good reason. Many measures are important to the business or organization. We are not abandoning those measures, the ones I call the little charts or sometimes called the dashboard. For Accelerated Continuous Improvement (ACI), we are choosing that single measure that will determine our success with our improvement initiative.

When the right measure (what one CEO called the pressure point—like on our body) is chosen, as it improves, you will see proportional improvement in related measures.

For example: When an aerospace company needed to focus on on-time program performance, the question was asked, "Is it possible to improve on-time program performance (OTPP) without increasing cost and decreasing quality?" They found that having chosen the right pinpoint, not only did it not harm other important key results, but just the opposite: It had a positive effect on them. On-time delivery improved by 35% while cost performance improved 37%, and both results were made possible through improved quality.

Consider these one-measure examples from departments at a greeting card manufacturing factory:

- **Time to Efficiency (TTE)**—When starting up five new sorters that were completely changing operations in the factory, the number of days to start up was measured.

- **Idle Paper**—Number of days from receiving a roll of paper until it is converted into product and shipped.

- **Monday Morning Lights**—After the weekend shutdown, how many hours until the factory is up and running at scheduled capacity?

- **Hours to Impress**—Time required to complete all impressions in a print run.

- **Picnic Table-Cover Waste**—How much of a football field would this week's waste cover?

Goal Setting

As part of the task of developing the measurement and feedback systems, consideration should be given to the issue of goal setting. These goals that will become important in designing, planning, and executing recognition and celebrations that will be discussed in Chapter 7. Setting goals for improvement of 5% or 10% is a bit of nonsense. It would be fair to ask, "Where did those numbers come from? Did someone just make them up—pull them out of the air?" Instead consider these principles and good practices for goal setting.

Goal Principles:

- Goals and targets are activators for improved performance (i.e., they prompt desired behaviors).
- The purpose of a goal is to provide opportunities for reinforcement.
- Think about goals as "milestones" along the way that are worthy of pausing and recognizing—not a final destination.
- Goals should be attainable, challenging, and rally-able—something associates can be proud of.
- Always, always have action plans for goals. As Dr. Deming taught us, "Goals set ... without a method ... are a burlesque" (Deming, 2018, p. 64). Insist on action plans for every goal.
- Deadlines give people permission to wait.

Goal Good Practices:

- Use natural goals like "best evers." Many celebrations in sports, particularly the Olympics, are around establishing a new best ever.
- Longest "strings" or runs is another form of natural goals. Even when you reach 100% or the Super Bowl, the challenge then becomes to repeat, three-peat. To do it again, multiple times.
- Use baselines and best evers to set goals.
- If current performance is a long way from the desired level, then shaping (or intermediate goals) will be needed.
- State goals positively if possible.

Goals and Competition:

- Use goals against current self, past, and competitors.
- Do not compete against others within the organization.
- The question is not, "Am I better than my peers" but "Am I better than my previous self? Am *I* getting better?"
- The real competition is against others who provide the same product or service and are vying for market share. Lowe's® stores are not competing against each other to be the best Lowe's store. They are competing with The Home Depot® stores to be the best home-improvement stores. Internal competition will cause destructive behaviors.

Feedback Is the Breakfast of Champions (Let's Get Creative)

Pearson's Law states that "Where performance is measured, performance improves. Where performance is measured and reported, the rate of improvement accelerates."

The first great challenge of measurement and feedback for leaders is to answer the question, "What's on the y-axis?" The second great challenge is to "Make performance visible."

Make Performance Visible

Bicycle Computer

I'm an avid bicycle rider. As of this writing, I have ridden 9,420 miles on my Trek bicycle. (When I get to 10,000, I plan to buy myself a new bike seat.) The wireless computer on my bike tracks miles, time, and speed, and I'm constantly looking at it to see how I'm doing. Unfortunately, a few times the computer malfunctioned, or the sensor became misaligned. In those instances, I immediately had no interest in riding until the sensor was repaired. If I'm going to ride (and I do enjoy it), I want to get credit. The feedback fuels my desire and energy to ride.

The Black Curtain

Several years ago, to illustrate in workshops the importance of feedback, we rented a bowling alley and hired a group of actors to produce a film in which a black curtain hung across a bowling lane, and you could not see the pins. As you can imagine, the bowlers did not have much enthusiasm for rolling the ball down a wooden lane and watching it disappear behind the curtain, sometimes hearing a crashing sound.

I would suggest that many organizations today have that exact setup. They are asking associates to roll the ball without knowing the results. The associates roll the ball, and it disappears behind the curtain, sometimes with a crash. In the video, the supervisor is behind the curtain and from time to time comes out and says something like, "I don't know what you're doing out here, but it looks bad behind the curtain." As the supervisors are taught the importance of feedback, they might come out after a frame and hold up four fingers, meaning, "You missed four!" (instead of "You got six!"). Over time, the bowler figures out that there must be 10 somethings behind the curtain, because that is the most you can miss. And over time, they are saying, "Pay me enough money, and I will roll the ball all day long. Just don't ask me to get excited about it."

As leaders, we must tear down the curtain so that everyone can see the target and the results. We must set up the pins, teach better methods for rolling the ball, keep score, and recognize improvement.

Bowling for Accelerating Improvement

- **Set up the pins:** Clear, understandable, and measurable goals. It is important to know what is expected.
- **Roll the ball:** Actions to eliminate problems and drive good practices.
- **Keep score:** Immediate, simple, objective data about performance. Measure and provide feedback.
- **Celebrate (reinforce):** Something in the job environment that recognizes good actions and results. Celebrate progress and reinforce results.

Timely & Specific Feedback

Useful feedback is information about past performance that is *timely* enough and *specific* enough to allow performance to change.

In other words, feedback *during* the game—while there is time to do something about it. This is not reading the results of the game in the newspaper the next day or seeing it on the postgame show. This is also not the information in the monthly accounting report.

Event-Oriented

When does the score get updated in a baseball game (basketball game, football game)? Answer: As soon as the runner touches home plate (the ball goes through the hoop or over the goal line). The scoreboard operator does not wait until the end of the inning to update the score. Our challenge is to develop feedback like this.

The absolute best feedback (and reinforcement) occurs "while sweat is still on the brow." For example:

- The photos/videos/plays quarterbacks like Peyton Manning look at while their team is on defense. (The game is not over. The score can be changed before it gets reported in the media.)
- The pizza tracker that tells you where in the process your pizza is. (A similar "tracker" could be used for work orders, doctor visits, or lab results.)

If the measure is to be time-based, and many will, shift is better than day, day is better than week, and week is better than month.

If at the end of the day/shift/week people are not excited, anxious to see the score, we don't have a feedback system that is timely enough, specific enough, and important enough to drive improvement.

Creative Feedback

CASE FILE

Feet of Wrapping Paper

With the pinpoint "feet of wrapping paper produced per dollar spent" displayed at the entrance to the factory where all workers walk in, lay four colorful sheets of wrapping paper on the floor for the workers to walk in and out on each day.

- One sheet represents last year or the baseline.
- One sheet represents this year's goal.
- One shows last week.
- One shows current year-to-date performance.

Each sheet is the actual length of the number of feet being represented (approximately 100 feet). The last week and current year-to-date performance sheets are updated each week.

CASE FILE

Machine Uptime—On Track

What better way to make uptime performance (machines, equipment, computers) visible than with a NASCAR-like racetrack, including a pit row. The objective is clear: Keep the cars (machines, orders) on the track and out of the pits. When a machine goes down:

1. Repair it as quickly as possible (measured as TTR: Time to Repair).
2. Repair it such that it does not fail again (measured as TBF: Time Between Failures).

Utilize a Hot Wheels® racetrack and numbered cars to represent each piece of equipment or each order.

CASE FILE

The Sales-Call Scorecard—Playing One-on-One With the Customer

Calling on a customer is somewhat like playing a round of golf in that it takes about 4 hours, requires preparation to get things ready, and follow-through to put things away. Given that analogy, we gathered top sales representatives from around the globe at Newport Beach, California, for 3 days to develop our sales-call scorecard (see Figure 6.2).

After literally covering the walls of the conference room with "What does it take to have a successful sales call?" and then boiling the ideas down, we arrived at the 18 elements of a successful call. These included a front-nine holes of what the sales rep should do and a back-nine of what we wanted the customer to do (since the overall objective of the

Sales-Call Scorecard

Customer _____
Date _____

Call
Score [ ]

Sales Representative

Hole	Par	Score
1	5	—
2	5	—
3	4	—
4	3	—
5	4	—
6	4	—
7	4	—
8	4	—
9	3	—
	2	—
	2	—

1. Present solution to customer problem
2. Ask for increased business or order rate increase
3. Propose "next step" in selling/buying process
4. Learn about new or key contacts
5. Propose entertainment
6. Learn customer needs
7. Reinforce customer
8. Follow-up from previous call
9. Sell company/self
   Bonus: Follow-up on CSS
   Bonus: Make joint call with invited reps

Rep Score [ ]

Customer

Hole	Par	Score
10	5	—
11	5	—
12	4	—
13	4	—
14	4	—
15	4	—
16	4	—
17	3	—
18	3	—
	2	—
	2	—

1. Works with you to solve problem
2. Agrees to place initial order or increase existing business
3. Agrees to "next step" in selling/buying process
4. Share information on key contacts
5. Agrees to entertainment
6. Shares current company needs & competitive information
7. Customer shares forecast for future company material requirements
8. Discusses their customers' products, markets, & concerns
9. Shares personal invormation
   Bonus: Invites you to join them in a personal event
   Bonus: Completes and returns customer satisfaction survey on time
   Bonus: Reinforces company or gives quality award

Customer Score [ ]

Figure 6.2. *The sales-call scorecard provided the checklist for a successful customer visit.*

call is to "change the customer's behavior"). In addition, since people like bonuses so much, some bonus items were added for both the front-nine and the back-nine.

For example: Sales Rep Hole #1, par 5, is "Present a solution to a customer problem." Customer Hole #7, par 4, is "Customer shares forecast for future company material requirements." A bonus example is a customer invites sales rep to join them at a social event, par 2.

Each sales call is scored, and opportunities for improvement are identified. The overall team score is tabulated and plotted. Over time, the holes and pars may be changed based on usefulness.

Note: For this "golf game," if you complete the task, you score par and ... the higher the score, the better.

As the sales-call scorecard is implemented worldwide, teams celebrate progress with meals that include ParBQ (barbecue), greens (turnip), and tee (sweet tea). They also play virtual golf and trade golf trading cards autographed by champion golfers.

CASE FILE

Twins' Bedtime Checklist

There is no greater blessing than a grandchild. Make it two at a time and you have twins Ava and Ivy, our first grandchildren. They were a barrel of fun and joy most of the time, but some tasks were doubly challenging with two at once.

When they were about 3 years old and my wife Debby and I were enjoying our overnight "babysitting" time with them, the bedtime routine was one of those challenges. Trying to get the two of them undressed, bathed, hair dried and combed, teeth brushed, pottied, and pj's on was somewhat of a circus ... until I developed the bedtime checklist (see Figure 6.3).

It worked like this: one sheet of paper for Ivy and one for Ava. Since the twins could not yet read, pictures were used for the items to be completed before bedtime. A sheet was given to each girl, along with a crayon. Their task: Complete each item, check if off with the crayon, turn the sheet in to Grandmother or Granddaddy, and you will have three books read to you before being tucked in. Problem solved! They took the sheets and set right to work to complete the tasks—and even enjoyed it. It also gave them the feeling of accomplishment and independence.

⚡ KEY POINT

This example illustrates the power of a checklist. I know of no more powerful tool for shaping performance. Make performance visible!

The concept of contingency is clearly demonstrated here as "if/then." If (or when) you finish the checklist, then you get to have three books read to you. Contingency puts the performer in control. It's up to them. Complete the list and get books read to you. It doesn't really matter to Granddaddy; the choice is theirs.

📑 CASE FILE

Million Dollar Pyramid

What does it take to sell a $1 million piece of equipment to a hospital? The equipment had been in use in the U.S. for some time, but there had been little market penetration in the Asia-Pacific Region. So, a 3-day workshop with the top medical-equipment technical sales reps was conducted. Flip charts and walls were again covered from floor to ceiling with ideas of what it takes to get the order. The multitude of ideas was reduced to the vital few. It was felt that a pyramid and a climb represented the challenge well. So, the essential elements were arranged in the pyramid based on degree of difficulty and progress over time (see Figure 6.4).

For example, on the base level, 10 points would be earned for acquiring a copy of the hospital's organization chart, and 20 points would be earned

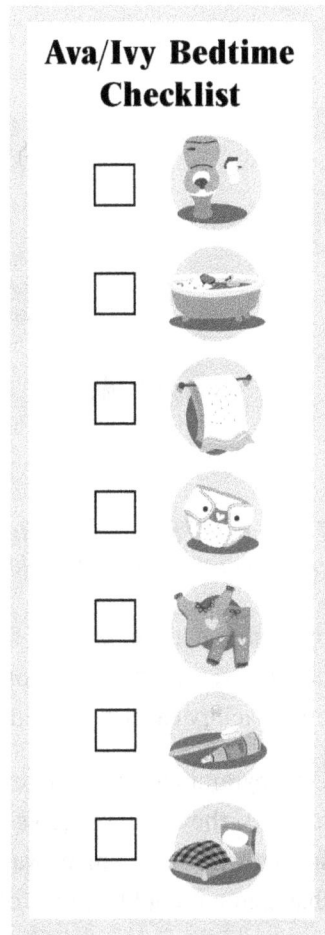

Ava/Ivy Bedtime Checklist

Figure 6.3. The twins' bedtime checklist greatly simplified the nightly ritual.

Figure 6.4. This diagram illustrates six examples from the dozens of elements in the Million Dollar Pyramid.

by finding out the name of the person who would sign the purchase requisition. A visit by the customer to see the equipment in operation at another facility earned 80 points.

On a pyramid for each target customer, the name of the hospital and the name of the technical sales rep were listed. The pyramid was designed so that when an element was accomplished/completed, the covering for that element could be torn off and discarded, changing the color of that element from yellow to red and initiating a celebration. A scale along the side of the pyramid showed the number of points earned against the possible total of 1,540 points.

We quickly realized that when the total score reached about 1,200, we could have 80% confidence that the order would be placed. We had taken the mystery and magic out of the sales process. To get the order, do the things on the pyramid!

Note: I quickly realized that a pyramid like this could be used to accomplish any goal.

For help in developing your pyramid to success, see the contact information in Next Steps—Available From the Author at the end of the book.

Cows on the Sick Lot

On the dairy farm, it's inevitable that cows will get sick from time to time. And when they do, they are "out of production." But ... an unbelievable number of things can be done to keep them from getting sick and to help them get well as quickly as possible. The next question soon becomes, "How do we get the farm workers to pay close attention to these important items?"

One answer is the Offline Medical Center sign, which is a simple display of the number of cows on the sick lot. Once the sign was put up, workers checked the sign often, talked about why the cows were there, what could be done to prevent it, and how to get them back online as quickly as possible.

Rotten Tomatoes—Downtime Reduction

In Ripley, Tennessee, where you find the famous Ripley tomatoes growing, Tomato World, and the Tomato Festival, what better way to represent a machine that is down than as a "Rotten Tomato"? And what better way to make performance visible than with a 10-foot wooden tomato-plant model?

The ACI plan called for machines scheduled for today to be represented by a ripe, red tomato on the wooden tomato plant. As a machine goes down, the healthy, red tomato would be flipped over to the rotten-tomato side. For associates passing by the large tomato-plant display, the rotten tomatoes (machines down) could be easily seen. On the wall by the tomato plant, the counts for machines down today, this week, year-to-date average week, last year's average, and best week ever make performance visible.

Reinforcement ideas included tomato sandwiches, biscuits with tomato gravy, fried green tomatoes, and homemade tomato soup.

How Low Can You Go—Limbo

Symbolize reducing downtime on equipment with a limbo stick (see Figure 6.5). The downtime scale is shown on the supporting poles, and the limbo stick is moved to the appropriate downtime level each day to track performance. Daily, week-to-date, year-to-date, last year, and best-week-ever downtimes posted on the wall beside the limbo stick give the overall picture. Limbo contests make for great reinforcement.

✍ THE STORY

I'll Never Do Another Contracting Job Without Flip Charts on the Wall

We lived out in the country on Buttermilk Road. Our dream house, designed by the area's premier architect, Uwe Rothe, was a four-story contemporary, with cedar siding and beams. We overlooked the valley with a view of 50 miles to the west. The house sat on 3 wooded, steep acres with

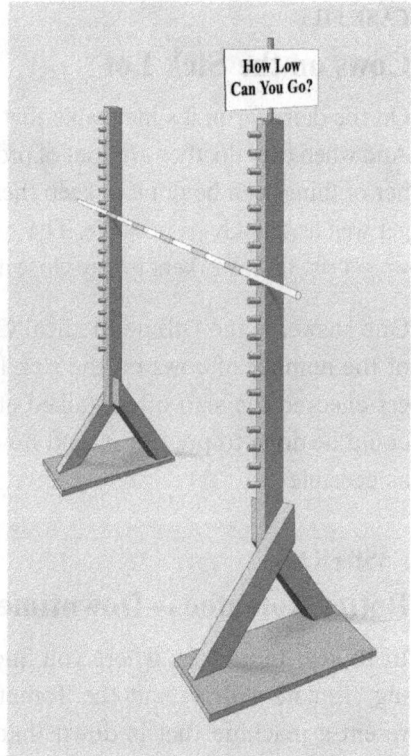

Figure 6.5. The limbo stick provided the perfect scoreboard for downtime.

no lawn to mow, only trees. Our family and friends loved the place. It was a bit of a showcase home—a favorite place for sleepovers, youth gatherings, Sunday School parties, and team meetings for work when we needed a creative place to get away.

But ... I was traveling a good bit internationally at the time, sometimes gone for a week or two, and the house was out in the country. We had one neighbor next door, but the next house in each direction was a quarter mile away.

Coming home from work one day, wife Debby announced that she wanted to move. She loved the place but did not like coming home at night when I was out of town and having to walk into a house that had no garage. She wanted us to move to a neighborhood. So, we began looking for neighborhoods and lots that we liked and soon found one on Eric Court.

Still though, it was a painful thought that we were leaving this place that we loved so much—one custom designed for us and one we had even helped build, even nailing the tongue-and-groove flooring and sewing the cushions for the kiva (conversation pit), for example.

It should have been one of my first thoughts: Why not build a garage onto the house? Maybe it wasn't so easy on the side of the ridge, but probably possible. Debby liked the idea too because she really loved the place. When I asked an architect friend about the garage, he said, "You can build anything you want. It's just a matter of the cost." I asked him to take a shot at designing it, and the price seemed reasonable. But, as the family began learning about this plan for a garage, other ideas started popping up.

- Debby thought that while we were in the middle of a building project anyway, why not convert the screened-in porch out back into a sunroom (something she had always wanted) and build a new screened-in porch?
- And how about that bathroom that was planned for the top floor but never finished?
- Daughter Lauren liked the idea of a playroom just off her bedroom above that new sunroom.
- Son Matthew thought that surely now was the time to complete the first floor and make it into a teenage hangout with a kitchen area and a deck.

And by now, I was getting into it and decided I needed a new deck out in the woods that would be my special place, one connected to the house by a walking bridge.

Checklists Save the Day

Needless to say, this new "garage" was going to end up costing me a fortune. Yet, the job began. And each day at lunch, I went home (a 15-minute drive) to check on the progress. As the Captain of the World, I would see things that needed attention. Items like:

- the trim under the sink in the new teen hangout/recreation room
- the door to the new playroom

- blinds for the recreation room
- pegboard wall in the garage
- door to the attic

When I pointed out these items to Ray, my contractor, his answer was always the same, "No problem, Mr. Justice. We will get right on it." However, days passed without my seeing any of the items taken care of. With growing frustration, I devised a simple feedback system. I got a flip chart like we used at work and posted it in the garage where Ray and all the workers could see it (see Figure 6.6). I listed all the "to be taken care of" items on the sheet with a box beside each one to be checked off when completed.

Figure 6.6. The carpenter's checklist, after surviving some damage.

The next day at lunch, I returned to the job site, found Ray, walked with him to the place where the scoreboard was posted and reviewed the list. Together we checked off all the items that had been taken care of. This continued for several days. One day when I arrived to go over the list, Ray told me that he had already gone over the list and checked off the completed items and that I could go see it for myself. He had taken over responsibility for managing the scoreboard.

Within a week, the workers began checking off the boxes as they completed items, not waiting for Ray to ask about them or letting him have the pleasure of checking the box and stealing their credit.

Over the next couple of weeks, I expanded the system to include a sheet for each group of workers—carpenters, plumbers, sheet rockers, heating/air conditioning, and so on. I was still reviewing the sheets with Ray a couple of times a week. Then one day at lunch, when I went to look at the scoreboard, I found new items added to the lists written in someone else's handwriting. When I asked Ray about this, he said, "Oh, the plumber started putting things on the carpenters' list, and now they are all adding items when they need help from each other."

Feedback Is the Breakfast of Champions

The crowning moment came when I noticed the plumber coming down three flights of stairs one day, from the top floor to the basement, then immediately climbing all the way back to the top floor, carrying nothing either up or down. Curious, I went upstairs and asked him the purpose of his trip. He smiled and said, "I went to check something off my list."

Not knowing in the beginning where this flip chart system would go, I realized we had arrived. A system that provided the feedback needed to help the team get the job done with simple, clearly defined expectations and immediate, encouraging feedback. As we wrapped up this project, Ray said to me, "I will never do another contracting project without flip charts on the wall."

For me, I realized that I know of no other management tool that is more powerful than a checklist: a living, real-time checklist with items added as needed (expectations defined) and items checked off when completed (credit given). And, by the way, no need to fuss about the items not completed yet. They are not going away. They have been made visible and keep staring us in the face. Focus on the completed work, give credit, and move on to the next day.

PS: The home on Buttermilk Ridge in East Tennessee was one day sold to a couple from Connecticut as "A Million Dollar View for Only a Quarter Million Dollars."

My Favorite Scoreboard

RJ was a Navy captain before he took over as the supervisor of Warehousing Operations at our Tennessee facility. You can be sure he ran a tight ship. So, you can imagine his angst when bale damage cost rose by 75% over a few months—from an average of 40 bales damaged per week (the "standard") to around 70. It had his full attention, and that attention led to him calling on my group, Engineering Services, for some help. His charge to us: "Get bale damage back down where it belongs." Rod, an excited young engineer, was assigned to the project and began to investigate possible causes of the increase, looking at possibilities like workload, new employees, press cycle, packaging supplies, fork-truck design, and row spacing.

In the meantime, there was a need to provide feedback to the workforce on this increase in damage. Up until now, bale damage was something that the supervisor kept track of and the workers had to fix. A monthly report of bale damage went to the supervisor for review.

A feedback system was designed that included the following:

- A card, with a diagram of a bale on it, to be filled out each time a bale was damaged in handling or discovered damaged when being staged or loaded on the truck.
- The location of the damage on the bale was noted—corner, strap, hole in the wrap—along with the possible cause.
- There was no place on the card for the name of the operator. The purpose of the card was not to assign blame, but to identify the cause and then find ways to prevent it.

The card was then dropped in a box (like a ballot box). Every 4 hours, the cards were removed from the box and counted. The count was used to update the daily and weekly scores on a scoreboard (see Figure 6.7). The cards were then collected to be studied by the workers at a weekly meeting.

The scoreboard was a 4' x 8' sheet of plywood with slots for today, this week, year-to-date weekly average, and weekly goal.

To make a long story short(er), the cause of the increase in damage was identified—a change in the cardboard wrap used on the bale from

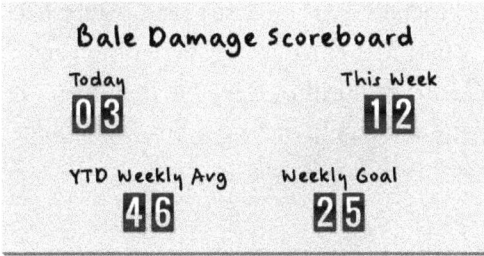

Figure 6.7. A 4' x 8' sheet of plywood served as the bale damage scoreboard.

a double-wall wrap to a single-wall wrap—which had been tested and determined by materials engineers to be strong enough and cheaper.

However, by the time the cause was identified, the damage had been reduced, not only back down to where it belonged (at 40), but to a goal of 25 or less per week while still using the new, single-wall cardboard.

The workers had learned the care that was needed to handle this more delicate bale. They had also made improvements in straight rows, row spacing, clamp design, clamp pressure, and other factors.

Over the next 6 years, the weekly average bale damage was reduced from 65 to 5 (see Figure 6.8). Beyond that, the system remained in place. A new box was added to the scoreboard for "Best Week Ever." And the weekly average continued to decrease.

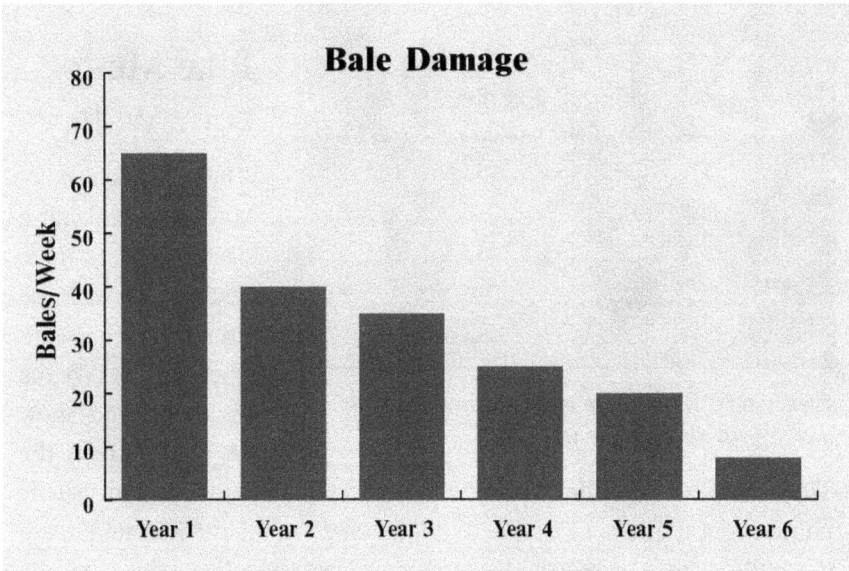

Figure 6.8. Bale damage steadily decreased over the next 6 years.

When the best-week-ever score reached *zero* one week (see Figure 6.9), the question was asked, "Now what? You can't get any better than zero." Arriving in the warehouse one Monday morning, RJ found that the scoreboard had been redesigned by the workers to include a box for "number of weeks in a row at zero."

Observations

This is, by far, my favorite scoreboard design of all time (for a manufacturing/production area): a 4' x 8' sheet of plywood. For office environments, a similar simple scoreboard can be posted on a wall and updated by a person who works in the area. This may involve hanging numbers, putting data points on a graph, or drawing lines between points. Such a scoreboard is nothing like the beautiful, multicolored, computer-generated graphs under plastic that I call "charts for the boss." (See Take Me to Your Walls below.)

This story points out once again that natural goals like best ever and longest string are far more rally-able than those set by management. (See the discussion earlier in this chapter about goals.)

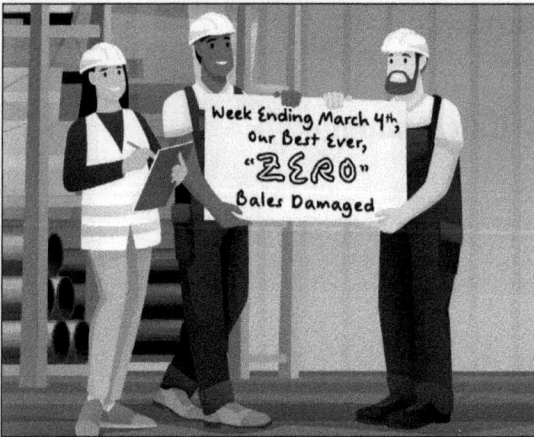

Figure 6.9. Warehouse associates celebrating their zero bale damage week.

Take Me to Your Walls

When I visit a company or organization for the first time, they want to walk me through their work process, then take me for a tour and show me what they do. They want me to understand the steps of their operation. I'm usually polite about it and act interested. But ... what I'm really interested in are the walls. I want to see the leadership process, which is evident by the feedback on the walls.

I believe I can walk through a factory, office, hospital, school, warehouse, restaurant, theater, or other organization, take a look at the walls, and have a rather good idea about the potential for improvement. If I don't see evidence of performance being made visible, I'm sure there is huge potential for improving.

Once while on the tour in a factory in Australia, I saw a chart that interested me. So I asked the supervisor of the area who was giving me the tour to explain the chart. His response, "If you want to understand that chart, we will have to get someone from accounting out here." In other words, these were charts for the boss. Such charts are colorful, updated by Accounting, and accurate. However, they are of no value for feedback, learning, and improving. Don't confuse data, information, charts, and graphs with feedback. Feedback is information about past performance that is timely enough and specific enough to allow performance to change.

Later, during that same tour of the factory in Australia, we came across two charts side by side. One was a chart on monthly production. This was April, and the chart was last updated in February. Next to that chart was another one with the score and statistics from the previous night's cricket score. Which topic do you think the employees in this work area were talking about—production or cricket?

The Four Questions

Say I was on the tour and asked the supervisor how their unit was doing. They said, "Follow me."

They took me over to a wall with two charts on it; one that said "The Big Picture" and another that said "Our Picture."

They pointed out that the Big Picture was for the company and Our Picture for their work unit. They showed me how the Big Picture was on-time shipments for the company—last week and year-to-date—and how Our Picture showed their unit's performance on errors in shipping labels for the same time periods. I would probably thank the supervisor for the tour, wish them good luck, and say goodbye. I'm not sure I could help them very much.

That supervisor and that organization are answering the four key questions that every employee should be able to answer on the spot:

- What is the organization focused on?
- How is the organization doing right now, today?
- How does my unit impact that organization focus?
- How is my unit doing right now, today?

The Walk-By Test

To be effective, scoreboards must pass the walk-by test. That is, you must be able to walk by the board, take a look as you pass by, and tell today's score to the next person you meet. If you have to stop in front of the board and study it or ask someone to explain it to you, it does not pass. It is not a scoreboard.

From Competition to Cooperation

Move away from competition between work groups to cooperation by emphasizing daily measures and scores instead of shift scores. If the graph on the scoreboard is measuring daily results, then on the scoreboard (beside the graph) should be boxes for each shift to record their score and a box to sum up the total for the day. The total for the day is plotted on the graph. The emphasis is not what a shift crew did, but on the day. One shift may have a lower score but be setting the next shift up for real success. The daily score will be permanently recorded on the graph while the shift scores (not individually important) will be erased to start a new day.

In the same fashion, if performance on the graph is being measured weekly (often the case), as shown in the "ready to roll" example in Figure 6.10, there would be a log for daily and week-to-date scores. This allows for variation between days, as a weekly target is being aimed for. On Wednesday, you might hear comments like, "If we keep this up for 2 more days, we can set an all-time weekly record."

Note also on the scoreboard boxes for the baseline, year-to-date average, and best week ever. Note also that the scoreboard is not computer generated, but hand drawn. I like it!

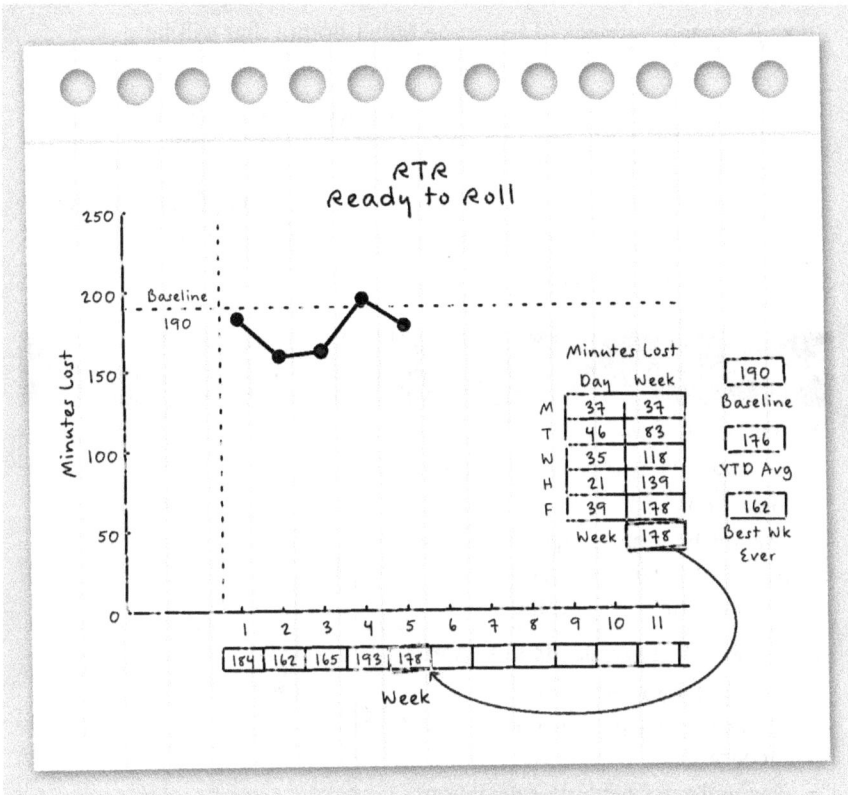

Figure 6.10. The ready to roll scoreboard showed daily, weekly, and year-to-date results.

My Accountability Scorecard

Each year for many years, in November I begin thinking about my priorities for the coming year. I ask myself what I really plan to concentrate on for next year. (I also begin asking myself what I have really accomplished in the current year and begin developing my list of highlights.) From that time of thinking them over, I develop my accountability scorecard (see Figure 6.11). You can think about it like signing up for a new semester of classes. It's a new semester and time to make some choices.

- What subjects am I going to sign up for, commit to? When I pay my money, there's no backing out.

- I also add some bonus items, not committing to them but wanting to pursue them.

- For each "subject," I list some bullet points that will help determine how well I am doing in that subject.
- Then, on a quarterly basis, I grade myself on each subject as fail, pass, or pass with honors.
- After grading myself, I sit down with my accountability partners and go over my "report card," sharing how I feel I am doing and getting their input.

Area/topic/course	Comments/ notes	Fail/pass/ honors
Listen to wife Debby • Stop other activity and pay attention • Move to be beside her as I listen • Use touch for most important messages • Plan Friday date days to provide dedicated time for listening		
SweetPea (Next grandchild to be born) • Pray daily for health • Gifts and notes to show involvement & encouragement • Plan and carry out birth celebration/ dedication • Explore options for 2nd residence in Tennessee		
Family • Weekly lunch with son Matthew • Grandgirls adventures, sleepovers, biking • Grissom family reunion • Update family notebook		
Ministering to managers/ The Transformation Network • Monthly Mr. Whiskers articles • Announce "This Is What Leaders Do" Workshops • Mentoring with Jay, Scott, Bob, Vic • Reestablish Mr. Whiskers web page		

Figure 6.11. *My accountability scorecard makes priorities and progress visible.*

Student Athletes

Have you ever thought about what you see when walking down the halls of a high school? I suggest at most places, it would be trophy cases and athletic awards on the walls.

Now, where do you think students put their discretionary effort? Where do you think their reinforcement comes from? Where do you think team-work and extra effort is learned?

What if? Just what if the hallways of high schools were covered with pho-tos and awards given to, say ... the best (highest class average) American History class that ever attended this high school (promoting continual improvement in the education process)? Maybe even recognizing honors classes, or those with 100% A's. Do you think we might have some aca-demic teamwork going on?

It's all in the design. The focus, the feedback, and the reinforcement. Both the academic and athletic systems are designed perfectly to produce exactly what they are getting. If we want to change performance, we must change the system.

Great leaders turn measurement into
feedback by making performance visible!

Have you ever noticed what is on the most visible (many times the only) bulletin boards in a company? It's often documents about government regulations and occasionally a "Think safety!" poster. (By the way, what in the world does "Think safety!" mean? It's a waste of good paper.)

Now for a couple more stories to illustrate the overall process.

📝 THE STORY
Flight to Delight

I worked with a Phoenix-based company that provided control systems for the aircraft industry. They had an unsatisfactory level of late delivery of systems and projects, so we launched the Flight to Delight (FTD) initiative with a kickoff ceremony attended by all employees. The kickoff included a presentation by Mike, the director of operations, who explained the challenge, asked for everyone's help, and committed to helping "release the champion" in each associate. The overall aim of this initiative was to eliminate delinquent orders.

The measure chosen was delinquencies per week. Beginning at a level over 1,000, they established 12 subgoals. Each of these subgoals was represented by a different city that would be "visited" on the FTD Tour. The cities where important customers were located included Memphis, London, Rome, Louisville, Tokyo, and Sydney. Plus, along the way were Boston, Philadelphia, Chicago, Wichita, El Paso, Seattle, San Francisco, Denver, and finally, returning to Phoenix. A large map displayed in the cafeteria showed the progress of the tour, and all employees were kept up-to-date through regular publications and team meetings.

Work groups in each operations area set aside time for working on reducing delinquencies as they continued with daily operations. Ninety-eight teams, each following the accelerating continuous improvement process, addressed and resolved production issues in work areas that prevented them from meeting delivery schedules to their customers. By the end of 18 months, 45 teams had met their initial goals.

The Champion Team, made up of representatives from each product line, led the FTD effort and met weekly to monitor the initiative's progress, make adjustments to plans, and make sure improvements were standardized and replicated. A Navigator Team was established to assist work teams with process improvements. The FACT Team published 16 newsletters full of team updates, pictures, link-ins, celebrations, and teaming tips. The Milestone Team planned all the best-ever and city celebrations. FTD purple shirts were worn on Wednesdays to maintain awareness. Pilot Perspective articles and regular updates were given in the *In-Flight News*. Adjustments were made to the initiative to include customer acceptance rates and final customer verification in the focus.

As the tour headed for the next city, the target level for that city and the highlights and features of that city were talked about to keep the next target in mind and to create anticipation for "reaching" that city. For example, when heading to Louisville, facts, photos, and customs of the

Kentucky Derby were emphasized. This "talk" also stimulated interest in identifying root causes of delinquencies and finding ways to eliminate them as permanent fixes.

Going from one city to the next sometimes took place quickly—as fast as 1 week apart. The tour landed in Boston, Philadelphia, and Chicago all in the month of November and in Wichita the first week of December. Sometimes the "flight" got off course because of "weather" or some other distraction (supplier-parts problems), and it took some extra effort to get back on track.

When each city was reached, the achievement was celebrated with city celebrations that focused on a theme for that city. For example, Memphis blues music, Chicago pizza, and a Boston tea party. Each city celebration incorporated a major customer from that city also. Of course, the celebrations included discussion of what was learned about decreasing late orders. Best-ever celebrations were conducted for new milestones in measures supporting on-time delivery. Continual emphasis was given to bringing FTD in for a landing back home in Phoenix.

After 2.5 years, the company had traveled across the country and around the world and returned home to Phoenix with on-time delivery averaging 98%. The 1,200 associates were welcomed back home to Phoenix with a blowout day of celebration that included a tent and band stage with entertainment, hot dogs, hamburgers, soft drinks, cotton candy, caramel apples, ice cream, a strolling magician, carnival games with prizes, dancing, a Velcro wall, Sumo wrestling, virtual gold, hoops, bungee jumping, bull riding, and so on. To top it all off, there was a flyover by planes that use their control systems.

By now, the company had become a star performer for their primary customer, to the point that the customer was so impressed they wanted to learn about the accelerating continuous improvement process.

All in all, the months-long initiative, the hard work, the creativity to identify and solve problems, the city celebrations at each milestone level, and the final blowout celebration brought great fun and pride to the company workforce. These were days never to be forgotten.

Material Effectiveness—PET Factory

As one of the earliest and biggest producers of PET plastics for use in the beverage industry, this company produced thousands of truckloads and hopper cars of plastic pellets each year to be sent to bottlers to be blown into soft drink and water bottles. The focus for the improvement that could contribute to company success best was readily identified as yield from the millions of dollars spent on raw materials. The choice was made to express the pinpoint as "material effectiveness"—how well the purchased raw materials were being used. It was thought that by using a new term, material effectiveness, instead of the traditional term of yield, attention could be drawn to the initiative.

Four kickoff meetings were held (one with each crew in the 24/7/365 operation). At each one, the four points of a kickoff were covered—each point by a different member of the leadership team. The scoreboard for tracking progress was introduced. Each work unit was asked to work together as a team to identify ways to link in to the initiative and to spend the time and employ their creativity to make improvements. With the help of continuous improvement coordinators, teams readily identified improvement opportunities and set to work on them. Before long, 40 teams in the plant were linked in and working on improvement ideas.

The material effectiveness scoreboard included a graph of "percentage right the first time" plotted each month. Also shown was percentage right for the past week and the month-to-date. The best-ever month and the year-to-date average were also shown. The scoreboard was updated each week on Monday so that the day-shift employees could see the results before going home. At the suggestion of an operator in the plant, a bar graph was added to show the equivalent number of trucks that could be filled with the product that was not right the first time. This tangible and visual way to represent performance was an attention-getter. A copy of the scoreboard was distributed to all work areas, to be posted alongside their project graphs.

Over the years, with this initiative as the focus, reinforcement opportunities abounded. Reinforcement actions included comments and notes on the scoreboard from the company president, and leadership team members meeting employees in the parking lot as they arrived for work (probably scaring them to death) to congratulate them on new performance records and progress. Reinforcement also included asking employees how they were linked in and what they were working on to bring about the needed improvement. At one point, when the plant reached a new best-ever period, the vending machines were turned on free for a day.

From an initial baseline around 90%, "right the first time" climbed to a level of over 99% over 7 years, steadily improving each year. Truckloads of "not right yet" decreased from a starting point of 142 trucks per month to one-fourth of one truck by the end of those 7 years!

Measurement & Feedback Summary

- Two great challenges:
    1. What's on the y-axis?
    2. Make performance visible!
- Measurement finalizes the pinpoint.
- If you can't measure it, you can't improve it.
- All measures are wrong, and some are useful.
- No data, no complaint.
- Measures go through three phases:
    1. You have a measure.
    2. You have an accurate measure.
    3. You have a useful measure.
- Feedback is the Breakfast of Champions.
- Good feedback is knowledge of results that is timely and specific enough to allow performance to change.
- Scoreboards must pass the walk-by test.
- Great leaders turn measurement into feedback by making performance visible.

Ready-Set-Go—Measurement & Feedback Checklist

As mentioned in Chapter 4, leadership teams should participate extensively in the design and feedback of the measurement and feedback system (scoreboard) as they begin to see this as one of the key elements for success and one of their key responsibilities as leaders.

☐ Identify that one measure that will best tell how much progress we are making in our initiative.

☐ Make sure the measure is expressed in the local (work area) language.

☐ To be impactful, the measure must be timely and specific enough to allow performance to change.

☐ You might need to try out several measures. Get historical data, plot on a graph, and take a look. Does this measure tell the story?

☐ Get started with your best idea for the measure. If it's not right, remember that the workforce will tell you and maybe even suggest a better one.

☐ Don't try to go straight to a design for the scoreboard. Have several different subteams sketch out an idea for the scoreboard. (A flip chart page or poster board is helpful.)

☐ As you hear the explanations for the scoreboard candidates and begin to understand them, it may become quickly obvious which is the "bell ringer." On the other hand, the ringer may result from a synergy of the ideas. In any case, when you see the ringer, you will know it.

☐ Once the scoreboard is designed, ask members of the workforce to take a look at it and explain it. If they are not able to explain the scoreboard, there is more work to do in the design (content and/or simplicity) of the scoreboard.

CHAPTER 7

Reinforce Behaviors & Celebrate Results

Let's Have Some Fun ... and Learn

Where does the energy come from to execute this improvement initiative, this strategy, this major improvement opportunity? Energy, enthusiasm, and creativity are absolutely necessary from all across the organization for such a major undertaking.

The answer is discretionary effort: The difference between the maximum amount of effort a person could bring to their task and the amount of effort required to get by (Daniels, 2000). In short, discretionary effort is the portion of one's effort over which a person has the greatest control.

In other words, this energy is the difference between what we do when we *want* to versus when we *have* to.

In behavioral science terms, discretionary effort is the difference in response to positive reinforcement (getting what you do want) versus negative reinforcement (avoiding what you don't want). In practical terms, this discretionary effort is the difference between compliance and commitment.

I have found through the years that this difference is incredibly large. Every associate brings to the job each day a pocketful of discretionary effort that they can spend if they choose to.

It should be noted that when I say "extra effort," we're not talking about working harder or being at work more. The emphasis is on the attention to detail while working—the desire for excellence (doing the job the best-known way) and a mindset for creativity, for improvement. Extra effort means constantly looking for a better way to get the job done. (Job #2 is finding a better way to do the job tomorrow.) Dr. Deming talked about the need for relentless, continuous improvement.

The driving force for this extra effort is positive reinforcement.

- Dr. Deming often said that recognition of some form is vital. He would say that you insult people by using only money. Instead, throw a party in their honor. Write a note on the board or graph. Let them train others to do what they have learned.

- Sam Walton, in his book *Made in America*, wrote "Nothing else can quite substitute for a few well-chosen, well-timed, sincere words of praise. They're absolutely free—and worth a fortune" (Walton, 1992, pp. 315–316).

- When asked at a conference how Milliken & Company won the 1989 Malcolm Baldrige National Quality Award, their VP of quality, Thomas J. Malone, spoke about applause as the secret.

- In a series of articles about his management journey, Jimmy Collins (2023), president and chief operating officer (COO) of Chick-fil-A®, talked about his efforts to reinvent himself from a critic into an encourager.

- "The things that get rewarded get done" is a central theme in Michael LeBoeuf's book, *The Greatest Management Principle in the World* (LeBoeuf, 1985, p. 9).

- Truett Cathy, founder of Chick-fil-A, said, "How do you know if somebody needs encouragement? If they are breathing" (Chick-fil-A, 2015).

- And my favorite of all comes from Dennis the Menace, who asks why he doesn't have a special place to sit when he does something nice?

Dennis asks a very important question. Most organizations have documents, procedures, even books filled with what to do when someone acts out. But where are the guidelines for how to recognize and reinforce individuals and teams for the right behaviors and good performance? It is one of the desires of my lifetime to see HR departments turned into "masters of reinforcement." Once that happens, if you need an idea for celebrating a best-ever or milestone accomplishment, go see HR; they are the geniuses of recognition and celebration.

The Four Levels of Reinforcement

There are four levels of reinforcement. All are good, appropriate, and worthy, but they have different impacts.

Level 1: Being nice to someone. Just recognize that people are there. Ask about their sick spouse or tell them you like the shirt they are wearing.

Level 2: Catch someone doing something right. For example, say to a co-worker, "That was impressive the way you handled that irate customer."

The next level is not Level 3; it is Level 10. Why? Level 10 is far more impactful than Level 1 or Level 2 reinforcement. Levels 1 and 2 are re-active—reacting to something that just happened. Level 10, on the other hand, is a proactive plan to drive improvement in a chosen area through a series of events and reinforcement—reinforcing behaviors, progress, mile-stones, and results, from kickoff to success. This is the level that brings Accelerated Continuous Improvement (ACI) a competitive advantage. This is the level we are designing for in this book. *This* is what leaders do.

The last level is Level 0, or it could even be expressed as Level -2 since it is not just neutral; it does more harm than good. This is where reinforce-ment is done insincerely or noncontingently. This is where the supervisor has a time slot on their calendar (Tuesdays at 2 p.m.) to go out into the work area and say good things about people. This is the leader who has been told that they needed to be more positive and say nice things. It would be better off for such leaders to just stay in the office and not be out setting fires everywhere they go by ticking people off.

Level 10 Reinforcement

Let me give you a simple example of Level 10 reinforcement—a series of reinforcing events throughout an improvement initiative. Let's say you are recognizing weeks in a row of 100% on-time delivery, no defects, no equipment stops, sales above last year, or all personnel needs filled. One way to do that would be by using symbols that represent the number of weeks in a row that the goal has been met. These symbols in themselves have no tangible value, but they represent the accomplishment, are fun, and create a memory. A symbolic plan might look like this:

- 2 weeks in a row–Doublemint® gum
- 3 weeks–Trident gum (or 3 Musketeers® candy bar)
- 4 weeks–a four-square game tournament
- 5 weeks–high fives for everyone (or 5th Avenue® candy bar)
- 6 weeks–six-pack of drinks (maybe add customized label)
- 7 weeks–7UP® soft drinks
- 8 weeks–Magic 8 Balls™ (custom messages inside)

There are those who would (and have) thought such items as silly or juvenile. It seems that it could be that way. But let me assure you; it is not. It is simply drawing attention to the improvement initiative and making work more fun. You will find that after about the third week when Trident or 3 Musketeers bars are given out, people will begin to say, "I wonder what will happen if we have 4 weeks in a row?" Boom! People are talking about performance and likely thinking of and discussing ways to make the success string continue. Discretionary effort!

Note: For the naysayers who make fun of this as silly, ignore them. Do not reinforce them for being negative, whining, and complaining. One of my most popular after-dinner talks, "Whine, Bark, Growl–Dealing With Difficult People" discusses how to handle these situations. To learn more about this skill, see Next Steps at the end of the book.

The Four Questions of Reinforcement

So, those are some of the key principles in this final component for accelerating continuous improvement, but what is the leader's job in reinforcing behaviors and celebrating results? Leadership must address four questions of reinforcement. If you answer them correctly, performance will accelerate to your expectations and beyond. The four questions of reinforcement are:

- *What* to reinforce?
- *Who* to reinforce?
- *When* to reinforce?
- *How* to reinforce?

Remember that Jimmy Collins (president and COO of Chick-fil-A) evolved from a critic to an encourager? It took him years to realize that the most important thing he could do in his role was provide encouragement. On his retirement day, after an extraordinary career at one of the most successful organizations ever, that was his conclusion. The most important thing he did was not determine the location of new stores, the introduction of new products, or the selection of the next officer ... but the encouragement of people. Don't wait until your retirement day to come to this important conclusion; start now to accept the responsibility as a leader to recognize and encourage your workforce.

Remember what Sam Walton said about sincere words of praise that are absolutely free and worth a fortune? Well ... they are not really free. It takes skill, creativity, and an investment of time to plan and deliver meaningful reinforcement. (We'll learn more later about the 4:1 rule.)

We will spend much of this chapter addressing the four questions of reinforcement: the principles and concepts, along with plenty of examples.

What to Reinforce?

1. Results *and* behaviors
2. Linked-in efforts
3. Progress toward the goal

Results *and* Behaviors

The question often arises, "Should we reinforce results *or* behaviors?" The answer is yes! We want to recognize both. Stated the best way, we want to celebrate the results and reinforce the behaviors leading to the results.

If we celebrate the results without making sure we understand the underlying behaviors, we may be making a big mistake. For example: Let's say as a supervisor, you are walking through a work area (factory, office, lab, kitchen) and see a graph that shows significant improvement, and you stop and write a note on the graph, "Good job!" There is a chance that you just made a mistake (Figure 7.1). Because you don't know if this is a good job or not until you ask what I call the magic question, "How did you do it?"

Instead of writing the "Good job!" note, along with a smiley face, find someone who works in the area and ask them about the chart. Ask them why it's important for this measure to improve and what has been done to drive the improvement. It's possible that you will hear something like this: "Well, things have sure been working well around here lately. It seems like things have just been coming together." What that really says is that the person doesn't really know why things are better. And there is no reason to believe "things" will stay this way.

Figure 7.1. You might be making a mistake.

If you hear an answer like that you should suggest that prayers be offered to thank God for improving things. But if actions have been taken to bring about the improvement, then write them on the chart so that everyone can see what caused the improvement, and it can be maintained. Then add your "good job" note and smiley face. Achieving good results and knowing how you did it gives you bragging rights.

CASE FILE

Thumbs-Up

In a leadership team year-end review meeting, it was pointed out that worldwide inventory had been reduced by $8 million as the year ended. After leaving that meeting, the company CEO passed Al, the head of Production Planning, in the hallway, thanked him for reducing the inventory, and gave him a thumbs-up.

On the way home that day while heading into the grocery store, I ran into Al in the parking lot. With great enthusiasm and his thumb in the air, he began telling me how Mr. CEO had given him and his production control teams the thumbs-up.

As I thought about the encounter later that evening, I began to wonder what the Production Planning department had done to reduce inventory. The next day, I told Mr. CEO about Al's excitement and his thumbs-up gesture. I then asked Mr. CEO, "How did they get that inventory down so much? What did they do?" Mr. CEO's reaction was a slap on his forehead, and he said, "I did it again, didn't I? I reinforced the event without finding out how it was accomplished."

At this point, Mr. CEO, who was well versed in the principles and application of reinforcement, realized that what he should have done was say to Al, "I saw that inventory was down by $8 million at the end of the year. How did you do it?"

Now, we all know that there is more than one way to reduce inventory. You can send out a memo to all the production planning and warehouse groups around the world and tell them to cut their inventory by 5% by the end of the year. If this was what Al did, then the appropriate response would be to remind Al that as a policy, we do not "cut" inventory—because we realize that in doing so, we may lose far more than $8 million in sales revenue as a result of stock outs and customer complaints.

On the other hand, what if Al had explained it this way? "Well, we worked with our suppliers to improve their on-time delivery. For example, this one supplier could be as late as 5 days and we had to carry a 5-day safety stock. Working with them, they improved to never more than 2 days late. That allowed us to reduce our safety stock by 3 days. We eliminated the need for that inventory."

Hearing an explanation like that—not cutting inventory but eliminating the need for inventory—is the occasion to get your hand out of your pocket for a good old thumbs-up for $8 million in inventory reduction, done the right way.

CASE FILE

Paper Mill Shutdown

When the line shuts down unexpectedly in a newsprint paper mill, it is a big deal—just like the production line itself is big. It's such a big deal that the mill manager wants to know about it and comes to the mill, even if it's in the middle of the night.

When the mill manager arrives, they will likely find a mechanic down under some part of the machine checking for the problem. Standing around will be a gaggle of people anxious about what the mechanic is finding out. This mechanic will be getting lots of attention from the supervisor and others asking if they need any parts, a better light, another tool, or any help.

If, in just a few minutes, they shout out that they have found the problem and fixed it, they will be a hero. The supervisor will be bragging about them, and the mill manager will be inviting them to go have a cold drink together.

Now what about tomorrow or the next day, when this same mechanic is under the machine doing routine oiling or checking the tension on a belt? They will not be getting much attention. If they had taken the day off, they might not even be missed. Yet these daily actions are what prevent the next big shutdown. You could understand if the mechanic didn't really mind so much if there was a problem that they had to fix. It seems that fixing problems is what they are paid to do. At least that's when they get attention, recognition, and hero status.

So, what to do about this? Checklists where daily actions are recorded help (as opposed to procedures written in a dusty book somewhere). One supervisor used an innovative idea: They placed handwritten notes in places where the mechanics were supposed to check on things like wear, corrosion, tightness, and so on. The note might be inside a panel that had to be opened, or the note might be taped to the wall near the work site. The note would read something like this: "Thanks for checking this machine and for completing the maintenance task. By keeping this piece of equipment running, we are able to produce over 300,000 feet of newsprint per hour." Sometimes the supervisor would leave it at that—a thank-you note. At other times, they would add to the note, "Please bring this note; come and see me. I would like to thank you personally."

When the mechanic comes, bringing the note, the supervisor would say, "Let's go back to where you found the note. I want to see how you take care of that piece of equipment." Once there, they would ask the mechanic to explain what they do and why it is important. The supervisor would add to the discussion the importance of the work and of the mechanic. They would thank the mechanic before leaving.

(It should be noted that if when arriving at the work area, the supervisor criticizes the housekeeping in the area or anything else, the mechanic will likely never "find" another one of the notes. Any problems found, unless they are of a safety nature, should be noted for future discussion.)

Linked-in Efforts

The second "what" to reinforce is linked-in efforts. Both the initiative to link in (which is celebrated in the link-in ceremony) and the progress of the effort should be acknowledged and reinforced.

CASE FILE

Show Me the Money

A woman picks up her daughter from soccer practice. On the way home, they drive by Pal's Sudden Service® (winner of the 2012 Malcolm Baldrige National Quality Award) as they do each day to pick up a strawberry milk-shake—her daughter's favorite. After she has placed her order at the order window, she realizes she does not have her wallet with her. Scrambling around the car for loose change, they come up with 92 cents. When she gets to the pickup window where she is to pay, she begins explaining that she only has 92 cents. The Pal's associate, without missing a beat, says to her, "Today is your lucky day! Strawberry shakes are on sale for 92 cents."

The woman says, "Thank you!" and of course she drives away with a smile on her face. But that's not all. The woman is a friend of Pal, the company founder and owner. She happens to know that he is out of town in New York. She phones him to tell him the story of the 92-cent milk-shake. He, in turn, calls the COO of the company and jokingly says, "I didn't know we were selling milkshakes for 92 cents now at the Stone Drive store." Pal then tells the COO the rest of the story.

Quickly taking advantage of a reinforcement opportunity, the COO calls the Stone Drive store and talks with Michelle, who sold the 92-cent shake. Perhaps she was worried that she was in trouble, but not for long. The COO tells her that he is proud of her for quick thinking and for putting emphasis on customer delight. The COO followed through by going to the Stone Drive store to celebrate the 92-cent shake, pointing out that while they may have lost money on that sale transaction, the

92-cent shake sale was linked to the strategic thrust of delighting customers, which was far more important.

Progress Toward the Goal

The last "what" to reinforce is progress toward the goal.

First Bicycle Ride

When daughter Lauren was ready to get her first real bicycle and learn to
ride, we headed out to the baseball field at a local elementary school. It
had a flat enough surface of dirt but wasn't too hard. I started her out on
home plate, helped her get her balance, and gave her a push to see how
far she could ride. She made it about 10 feet before falling over. I had a
stick with me, and I drew a line in the dirt marking how far she had made
it from home plate. I pointed out to Lauren that the line marked the fur-
thest distance she had ever ridden a bicycle; it was her "best ever," and we
would go back to home plate and try to set a new record.

Back at home plate, she got balanced again and took off with a push from
me. Each time that she rode further from home plate, I marked the distance with a line in the dirt. Soon she was riding around the baseball field.

Next, we worked on steering. I placed a quarter on the ground and told
Lauren it would be hers if she could run over it with her front tire. We did
this a few times until she had the knack of steering, and I was running out
of quarters. The visual markers made it easier for Lauren to succeed and
gave me more opportunities for reinforcement.

Archery

When son Matthew was a boy, we attended a father-son Royal Ambassadors Camp at Camp Carson in the foothills of the Great Smoky Mountains. Over the weekend, a series of activities was available to choose
from. One of those that Matthew was interested in was archery. But when
we got to the archery station, the target was much further away than a
10-year-old boy could shoot. I asked the teenage boy running the event
if we could move the target up for the weekend since all the boys at the
camp were 10 years old. With some disgust, he informed me that the target was at the official Olympic distance. Matthew and I moved on to the
next activity station.

What is the best way to become good at archery (or to enjoy it for that matter)? Do you start at the official Olympic distance? No! You start at a distance where you can hit the target, have success, and feel good about it. You can always move the target back—increase the distance—as you get better.

Shaping

The concept behind reinforcing progress toward the goal is known as *shaping*, and it is one of the cornerstone principles for accelerating continuous improvement. It amounts to reinforcing small steps or progress toward the ultimate goal. For example, if we are teaching a baby to walk, we don't send them to a class on walking or show them a video of how to walk. We don't wait until they are walking around the room to say, "Good job!" We help them stand up and we say, "Come to Daddy." If they take just one step, just one, we get on the phone and call Grandmother to tell her that little Lauren is walking. (For more about the concept of shaping, see Aubrey Daniels's [2016] book, *Bringing Out the Best in People*.)

As part of an initiative to reduce ink waste in a printing operation, we started with an excess-ink inventory of 9,000 gallons. The shaping goals established were 8,000, 7,000, 6,000, 5,000 ... down to 1,000, 500, 100. The numbers are easy to remember, and each level is worthy of recognition and talking about how we were doing it.

You may have heard the phrase, "Catch people doing something right" or maybe, "Help people reach their full potential; catch them doing something right" originating from management thought leader Ken Blanchard. The truth of the matter is that will not help them to reach their *full* potential. To help people reach their *full* potential, catch them doing something *approximately* right on the way to being right (e.g., baby walking, archery, bike riding).

Help people reach their "full" potential—catch them doing something approximately *right.*

CASE FILE

Film Across America

Film, produced at exponential speeds, is first wound onto large rolls and then sent to the splitter. While entering the splitter, the roles are inspected for defects. If a defect is observed, action must be taken—a very time-consuming and expensive effort. With a pinpoint of "defect-free film" (continuous production without a defect) and a measure of miles of film produced without a defect, the scoreboard could show cities reached as milestones of progress and celebrations tailored to symbolize the cites.

- For New York City, pizza.
- Buffalo, of course, buffalo wings.
- St. Louis—a six-pack of Budweiser®.
- Las Vegas—a pair of dice with a note to "not take any chances now."
- Hollywood—a weekend bash with free movies for the family at a reserved local theater.

CASE FILE

Five-Star Store

As part of the "world-class manager" initiative, a way was needed to recognize pizza store managers for their progress toward becoming a "five-star store." To provide a symbol of progress, criteria were developed for each star level (1-5) and communicated through a scoreboard.

Each month at regional meetings, store managers who had moved at least one star in a positive direction (e.g., from 2 stars to 3 stars) were recognized as "rising stars."

Each rising star shared the actions that were taken to move the store in the right direction, giving other store managers the chance to learn and replicate the improvements.

In addition, a star was added to the five-star store flag displayed in a prominent location in the store. As the star is added to the store's flag, a celebration with store associates takes place to acknowledge the achievement and to discuss how it was accomplished.

Who to Reinforce?

1. Those who are helping get the job done

2. Expand the winners' circle

Those Who Are Helping Get the Job Done

Reinforcement is not the job of *just* the supervisor.

It's everyone's job to reinforce those who are helping to get the work done.

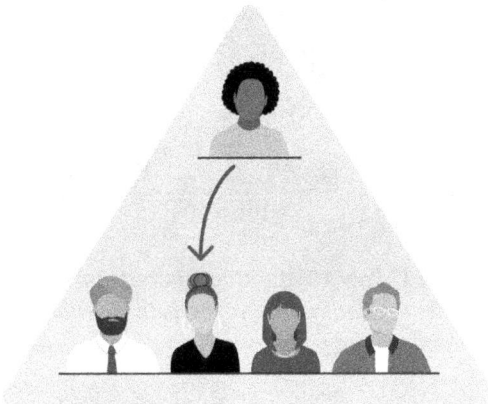

Figure 7.2. Traditionally, the supervisor was responsible for recognition and reinforcement.

In traditional thinking, we have placed almost the sole emphasis for the job of reinforcing on the boss to reinforce those in their organization at the appropriate time (see Figure 7.2). And, far too many times, that reinforcement has been directed at trying to get the underperforming associate to improve to an acceptable level. I suggest that is counterproductive. When the other associates see the underperforming associate getting reinforced for improvement, they will say, "What about me? I'm doing good work around here, but no one seems to care." You might even have some usually good employees acting out just to get attention.

Instead, I suggest establishing a measure of success for the *team*, along with a graph/scoreboard. Update the scoreboard once a day or week, and meet with the team in front of the scoreboard to discuss performance. When the measure shows improvement, ask the team the magic question: "How did we do this?" You will likely hear things like the following:

- "We could never have made this progress if Rick had not found the cause of the breakdowns."

- "Marsha found a way to stop the breakdowns from happening."

As the team discusses the actions taken, they will be reinforcing each other, and reinforcement will be bouncing around. You know the process is working when you begin to hear reinforcement directed back at the supervisor, like, "Yeah, and the boss got approval for that new instrument we needed" (see Figure 7.3).

So, what about that underperforming team member? Their name will not come up in the discussion of how we did it. And, I have found that over time they will either pick up the slack to do their part or they will leave.

Figure 7.3. Peer recognition where team members (including the supervisor) reinforce each other is the goal.

Now we are getting somewhere. We have a team with a measure of success that is improving, and we are meeting together to discuss how we did it and recognizing each other! I would suggest that peer reinforcement is even more valuable and powerful than supervisor reinforcement. In a sense, the supervisor "gets paid" to reinforce and may not fully appreciate the challenges and complexity of the job, whereas the peers know all that is involved in the job and the obstacles and hassles that must be overcome.

CASE FILE

Seminars in Europe

As part of a worldwide tour to offices in 15 countries to introduce the new process for improvement, my family, along with my workshops partner Nancy and her family, were in the Paris office. Now, everyone knows that the Paris office takes a measure of pride in rejecting new ideas from America. But at the end of the day, Nancy comes up to me and says, "You were on today. That is the best I have ever heard you tell the stories and explain the concepts. You even stole some of my stories and told them better than I ever have."

Now, when Nancy said that to me, my chest began to swell a little bit. That was real reinforcement, coming from someone who had stood in front of that challenging group herself and knew how hard it was to win them over (see Figure 7.4).

On the other hand, when I returned home after the tour and my supervisor, Frank, told me I did a "good job," I would thank him, but I would be thinking, "How do you know? You've never been in one of these overseas workshops. You've never stood in front of a group that speaks another language and is not too excited about this new process."

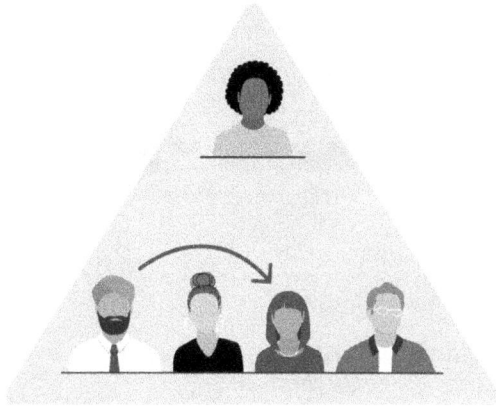

Figure 7.4. Extra special and very meaningful reinforcement from a work partner.

What I'm saying is that reinforcement from a peer is often more meaningful than reinforcement from the supervisor.

Beyond that, I believe there is an even higher, more powerful source for reinforcement: recognition from a customer—the person actually receiving the benefits of your work (see Figure 7.5).

On the same world tour, we conducted a workshop in Zug, Switzerland. A couple of weeks after we got back to the U.S., I received a note that went something like this: "I want to thank you for coming to Zug and sharing your approach to work and life. My wife and I have been going through a rough spot and while I was listening to you, I realized why. It was because of me. My critical nature and not appreciating her. We are working it out, and I believe we are going to be OK. Thanks for coming to Zug and sharing. You helped save our marriage. By the way, not only that, but I used to have to drag myself out of bed to go to work, but the office is better now."

Wow! Now that's what I call real reinforcement! You better stand back; I might explode from swelling up with pride and satisfaction.

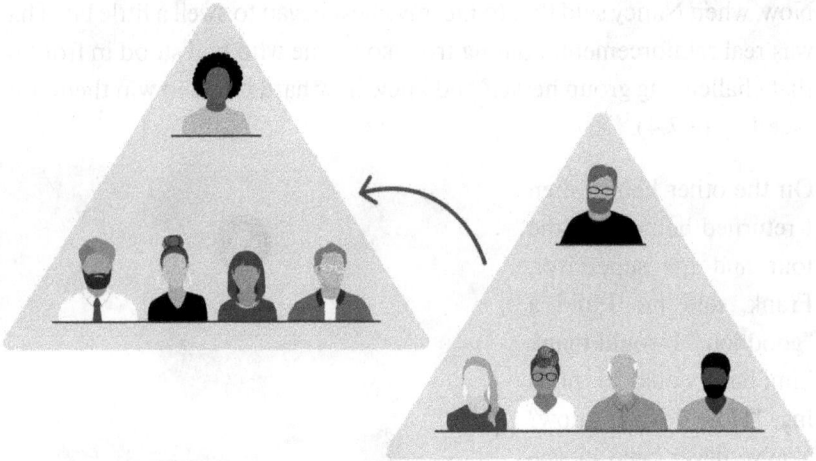

Figure 7.5. *Customer reinforcement is the most powerful because you made life better for them.*

I can only think of one scenario that could make that even better. If on a future trip to Zug with my family, that fellow had caught us having dinner by the lake one evening, come up to our table, and said those same things in front of my wife and children. Chest exploding!

Summary

Supervisor Frank's reinforcement is OK, but that's his job and he has no idea how hard my job is. Nancy's reinforcement is very meaningful because she understands the challenges. She admires what I have done. Reinforcement from a client who says, "I not only liked what you said, you changed my life!" is encouraging and energizing. It even makes me want to get on the plane and head out to Europe again.

Reinforcement is everybody's responsibility. It's everyone's job to recognize and reinforce those who are helping to get the job done, to make the job easier, and to make life better. That person may be a peer you work with every day, a supplier (someone who provides input to you), a customer (someone who receives your output), a supervisor, or even the president of the organization. When we have created an organization, team, and unit where reinforcement is bouncing around between associates, we will see improvement accelerating.

Reinforcement is everybody's responsibility. It's everyone's job to recognize and reinforce those who are helping to get the job done, to make the job easier, and to make life better.

I Was Helped By

One of my favorite reinforcement tools is the "I was helped by" board (see Figure 7.6). It works like this: A flip chart or whiteboard is placed in the work area (or on a comment screen on a team computer workspace) with the title "I Was Helped By" and plenty of white space below. During the day, if anyone helps you get your job done, you write the name of that person and what they did on the board. It only takes a moment and it's the right thing to do to acknowledge help received and say thank you in this simple way. (See the John Wooden story below.) At the end of a day or a week (depending on your situation), the chart is taken down (or board erased) and a new board is started. If you like, you can have a stand-up meeting around the board and discuss the names and actions (learning how to achieve excellence) before the chart is removed or board erased—simple, fast, and effective.

I was helped by:

Figure 7.6. Create daily or weekly reinforcement with a simple flip chart, whiteboard, or computer screen.

He Will Be Looking (John Wooden)

John Wooden, in my opinion the greatest coach ever, was known as a mild-mannered basketball coach who was a teacher to his players. The story goes that during a game, one of his University of California, Los

Angeles (UCLA), players stole the ball and headed down the court on a fast break. The player who stole the ball passed the ball to a teammate who was trailing close behind, and the teammate went in for a layup.

As the players were making their way back down the court, it was obvious that Coach Wooden was upset on the bench. As soon as UCLA got the ball back, Coach Wooden called timeout and brought the players over to the bench. He proceeded to chew out the player who had made the basket on the fast break. This was a rare event of seeing Wooden obviously upset, angry, and in the face of one of his players.

Wooden said to the player, "I don't want to ever see you do that again!" The player looked at him and asked, "Do what, Coach?" Wooden replied, "Get a pass from someone setting you up to make the basket, and not acknowledging the teammate who passed you the ball and set you up. All you need to do is point to him or nod to him on the way back down the court." The player replied, "But what if he is not looking at me on the way back down the court, Coach?" Wooden replied, "He will be looking!"

There are people all around us, helping us out every day in many ways *and* they are looking to see how we will respond.

📑 CASE FILE

It Looked Like the United Nations

Labels were falling off drums in the Singapore warehouse. When operators went to pull an order, they found labels on the floor next to the drums of product.

Singapore management was saying that those hillbillies in the factory in Tennessee can't even stick a label on a drum correctly. Stacker drivers in the Tennessee warehouse are asking how the operators in Singapore are causing the labels to fall off the drums.

The people in Tennessee knew (Singapore did not know) that warehouses around the world stored this same product, but only the Singapore warehouse was having a label problem. One of the stacker drivers in Tennessee figured out what was happening. Singapore is near the equator, meaning high moisture, temperature, and humidity. With this knowledge that the labels were not suitable for that environment, the label was redesigned for application in hot, humid environments.

The Singapore warehouse, so glad to not be dealing with this problem anymore, sent a small Singapore flag replica to the operator in Tennessee who made the suggestion.

I heard about this story and several people suggested I go down to the warehouse and check it out. Arriving in the warehouse, I did not know the stacker operator's name. So, I was going to have to find him. It didn't take long. Before I could even start asking, a stacker went zooming by me with a small Singapore flag taped to the front of his forklift. (I wish I had taken a photo of that for my archives.) I chased the operator down and asked if I could talk to him.

I asked if he was the person who had figured out the problem with the labels in Singapore. He said he was, and as he explained the history of how he had gotten the flag, he pulled a piece of paper out of his pocket to show to me. He said, "I got the flag, and this note from the people in the Singapore warehouse." Since he had gotten the flag and note days ago, I assumed he had been carrying it around with him since then.

The hero operator went on to tell me that one of his buddies in the warehouse wanted one of those Singapore flags. So, he sent a note to the warehouse in Singapore asking them to send him a flag. No surprise—they sent back a note that said, "Help us solve a problem and we will send you a flag." The buddy sent a second note to Singapore asking what he could do to solve a problem or make life easier for them.

It had begun. Tennessee warehouse supervisors sent notes to all the distribution warehouses around the world asking, "How can we help make life easier for you?" Not so long after that, the warehouse looked like the United Nations, as stacker drivers began to collect flags from different countries by solving logistics problems and improving the distribution process related to each country.

At this point, the warehouse had arrived! They had created a system of *want to* instead of *have to*.

Expand the Winners Circle: Employee of the Month

It has been a hobby of mine (encouraged by my friend Aubrey Daniels) for many years to study employee-of-the-month programs. If I go into a hotel, restaurant, or place of business with an employee-of-the-month plaque, I begin to get excited (and chuckle). It is a rare occasion indeed to find such a program that is working.

When I spot the plaque, I begin asking those working there about it. I ask the person in charge if I could talk to the person last listed as the employee of the month. Oftentimes the answer is, "They don't work here anymore." If I do get to talk to the person and ask them how they did it, what they did to become employee of the month, the answer often is, "Beats me. They just choose 'em somehow."

Usually, the program works like this. You start out, let's say, in January. At the end of the month, you ask the questions, "Who is the employee of the month? Who has shown the most initiative, produced the most results?" Usually there is a member of the team who is an exceptional, super employee, an easy selection. So, that person is chosen, and their name put on the board and recognized. Then month 2, February, rolls around. Who do you think will be the top performer in February? I'm betting it will be the super employee, the employee of the month for January. They were already the best and after getting recognized at the end of January, they turned it up a notch.

Wait a minute! We can't give the award to the same person 2 months in a row, so the question becomes, "Who is the next best?" Soon the question at the end of each month becomes, "Who has not gotten the award yet?" It has been my experience for over 40 years, around the globe, to find few employee-of-the-month boards with the same name listed more than once. In fact, many of the award boards never make it a full year. I have seen many organizations that just stop making the effort after a few months.

So, what to do?

EmployeeS of the Month

The answer is simple: add an "s" to employee, making it employees of the month. That way it is based on a set of criteria where anyone can win, depending on meeting or exceeding the goal.

CASE FILE

Hotels in Dallas

It started out as housekeeper of the month for each hotel. However, the program was not working well as there was complaining and bickering about who was chosen and how. The redesign consisted of creating a 1,000-point scale for the perfect cleaning and maintenance of a hotel room. The criteria and weights were designed with input from housekeepers, supervisors, and guests and included everything from the positioning of lamp shades to towel presentation. Each week, each housekeeper had one room audited at random. The audit was conducted by a supervisor and two housekeepers. (It was important to include the housekeepers in the audit because they probably learned more from conducting an audit than from receiving an audit report.) The scores for each housekeeper were tabulated for a month. Housekeepers with monthly scores of 700 or more were inducted into the 700 Club.

Then, the difference maker—if one housekeeper at this hotel was in the 700 Club, they received a $20 meal certificate at any of the hotel's restaurants. If two housekeepers at this hotel made the 700 Club, then each of them received a $25 meal certificate; five housekeepers in this hotel, a $40 certificate. Now, instead of competing with other housekeepers for housekeeper of the month (which led to secrets about how to score high), each housekeeper was cooperating with other housekeepers to not only get a high score for themselves, but to help others get a high score also. They had evolved from competitors to teammates. Housekeepers began identifying new items to be added to the audit sheet and volunteering to train new housekeepers. (Note: As we have said before, in both situations, the system was working perfectly just the way it was designed.)

Too Many Pink Cadillacs

While using pink Cadillacs as a method for rewarding associates for achieving sales targets, can you imagine the leadership for Mary Kay® saying, "We recognized too many associates with pink Cadillacs last year"? No. Absolutely not. Every time Mary Kay "gives away" a pink Cadillac, they are making money; the more winners, the better.

From Competition to Cooperation!

📋 CASE FILE

From Boss Award to Triple Crown

When Jim Fuller, my workshop partner, and I first started working with the warehousing organization, they had what they called the Boss Award—given to the warehouse in the U.S. that scored the highest on Boss Award criteria. The warehouses competed for the award each year.

No one really considered the impact of this competition, where warehouses did not cooperate with each other and kept secrets of good practices that they had developed and put in place. The warehouses were only acting logically. There is only one winner. Therefore, if I help you improve, I'm reducing my chances of winning.

We came up with a new design—The Triple Crown—that would be given to every warehouse that could meet the goals for three elements (Figure 7.7):

1. Shipping accuracy (the audit)
2. Shipping quality (the cube)
3. Shipping cycle time (how long it takes to get something from us)

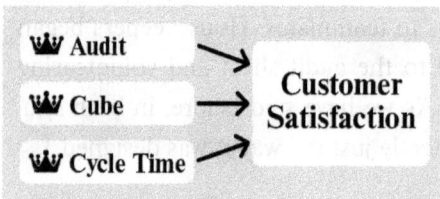

Figure 7.7. The Triple Crown award included accuracy, quality, and timeliness—all directed at customer satisfaction.

In addition, emphasis was placed on maximizing the number of warehouses that won the award. The more warehouses that won the award, the greater the recognition, thus creating cooperation between the warehouses to improve shipping accuracy, quality, and cycle time.

Each warehouse employee was now thinking, "I want to win the award, and I want to help all the other warehouses do the same."

CASE FILE

Car Salesperson of the Month

A family decides that they want to purchase a three-row SUV for their growing family, so they're off to the local car dealership. Walking onto the lot, they are greeted by the person who happens to be the *truck* sales expert, and they tell him that they are shopping around.

Now at this dealership they have a program for salesperson of the month (based on most sales dollars). After listening to the couple describe their interest in a new SUV, what do you think the salesperson will do? Will the salesperson try to fix them up with a new, three-row SUV or call for the person who is the three-row SUV expert? My money is on trying to sell them that SUV that will move the salesperson closer to being salesperson of the month.

What if, instead of salesperson of the month, this dealership had a scoreboard that plotted sales each day and showed month to date, year to date, best month ever, and best year ever? And anytime sales for a month exceeded the best month ever, each salesperson received a $500 bonus.

Now, what do you think? Would the truck specialist try to make the sale themselves or call the three-row SUV expert to help talk with these people? (Move away from competition to cooperation.)

Summary

In the housekeeper of the month or salesperson of the month examples, the person being recognized deserves to be. But the problem is that everyone else is saying, "How about me? I worked late one night last week to help us catch up. No one is acknowledging that." So ... we had been, and rightly so, making one person happy at the expense of making many others dissatisfied. But now we're adding the "s" to create many more winners.

Any system that has a limit to the number
of winners is a losing system.

📑 CASE FILE
X + 1

No, this is not a science fiction story. It's about a restaurant chain with more than 100 stores. It is also an organization with a store of the year award for the highest-scoring store on a set of performance criteria and a process that promotes competition within—destructive competition. The winning store's manager even wins a cruise for themselves and their spouse. As a result, the stores keep secrets from each other about improvements they are making.

If a store comes up with a way to reduce waste or improve speeds, there is no way that they are sharing that with the other stores. That would reduce their chance to be store of the year and for the store manager to win the cruise.

Fast forward to *X* + 1, a system that says that if 10 stores meet the world-class store standard, then each manager gets to attend the World-Class Store Banquet at The Lazy Lobster Fish Camp. If 20 stores qualify, it's an evening harbor cruise with dinner; for 30 or more stores, an overnight on the Delta Queen Riverboat. If 50 or more stores qualify, a 3-day Caribbean cruise is awarded to those store managers and their spouses. You get the picture—cooperation versus competition—just like in the Hotels in Dallas example.

(Note: In any *X* + 1 system, you have to make sure that the benefits of the improvement are more than covering the expense of the recognition.)

When to Reinforce?

1. When → then
2. Contingent
3. Earned

Really there is only one "when" to reinforce, but I like to state it three different ways. Perhaps the best way is "contingent": occurring or existing only if certain circumstances are the case; dependent on. "Earned" and "when-then" also capture the principle.

"People often tell me that motivation doesn't last,
and I tell them that bathing doesn't either.
That's why I recommend it daily."

—Widely attributed to Zig Ziglar

A Big Glass of Iced Tea

If you are out mowing your lawn and you say to yourself, "When I get to the walkway, I'm going to stop and get me a big glass of iced tea," you are practicing contingency—earning that glass of iced tea. When → then.

No Pass, No Play

Some years back, there was the movement called "no pass, no play" that said that high schoolers must be passing in school to participate in extra-curricular activities like sports, band, and so on. When this law was first passed in Texas, many expressed their concerns that it would result in more dropouts based on the belief that the only reason some of the boys came to school was so they could play football. And it turned out some of these boys were not passing their classes. But the Texas legislature held their ground. The result was that after a couple of seasons, the football players who in the past just could not pass (their courses, that is) were passing in all subjects and back playing football.

Stay in School—Drive a Car

The video clip on TV showed a highway running over rolling hills on a foggy morning. A yellow school bus topped the ridge, followed by a second bus and then a third. In all, 18 buses topped the ridge—all of them empty. The commentator said, "Every week, 18 busloads of students drop out of school." The action taken by the state was to make having a driver's license for 16- and 17-year-old students contingent on their being in school.

Token TV

You can buy one to attach to your television set—a token TV box. The TV will only turn on when a token is inserted, and then for a fixed amount of time. This device comes in handy with your children (and some adults). They earn TV time by finishing homework, completing chores, and so on. No, it's not bribery. It's a good lesson in life—a good lesson in the way the world works. After all, this is the same way most of us earn our paycheck by coming to work. Note that this is different from bribery, where you give someone the reward before they take the action.

Celebration Ticket

One of my favorite reinforcement methods involves celebration tickets. When a major accomplishment occurs and celebration is called for, tickets to the celebration are issued to all who contributed to the accomplishment with this note on the back: "For admission to the celebration, please write on the back of this ticket what you did to help us reach our goal." (Sometimes we change it up a bit by saying, "Write the name of someone who contributed to our success and what they did.")

The Way it Works—People enter the celebration and place their tickets in a box. After explaining what has been accomplished and why it is important, the celebration moves to the magic question, "How did we do it?" At this point you can say, "Let's find out." Pull some tickets from the box and read them aloud. It's enjoyable to hear the names and actions read out for all to enjoy.

Note: Some will complain about having to justify their attendance. But they should be kindly and quickly reminded that this celebration is for those who contributed to the accomplishment.

Contingency—The Story Cloth

Contingency is one of *the* most important principles for accelerating improvement and for accomplishing goals. It is the primary element of "when" to reinforce. That is, reinforcement must be earned. *When* you do something (behavior), *then* you receive the reinforcement. For example, when you finish mowing the lawn, then you get a glass of lemonade or get to go play golf. Examinations show that many so-called high achievers are zealous about managing their life this way. Everything is contingent for them. They design their life, schedule, and calendar so that they get to do the things they really enjoy when they finish tasks.

A few years back, I came across a brilliant and powerful example of contingency that beautifully illustrates the importance and power of the principle. A friend of mine, Pastor John Shepherd, led the Calvary Road Ministry. Part of that ministry is to the Maasai tribes in Kenya.

These tribes are organized into Bomas, each with an older man (Nyankusi) as the leader. By tradition, the Nyankusi will only listen to other older men, whom they consider their peers. Pastor John and other older men travel to the villages to minister to the Maasai and share the gospel with the Nyankusi.

The Maasai have a long-preserved culture where the language is spoken but not generally written. The strength of the oral tradition makes writing things down unnecessary. The Maasai history, culture, and traditions are passed from the old (primarily the Nyankusi) to the young through storytelling, songs, proverbs, or idiomatic sayings.

As a tool to communicate, a story cloth was developed (McAlister, n.d.; Figure 7.8). Each picture tells the story of an important event from the Bible. When John and his coworkers shared with the Nyankusi the story behind one of the pictures, at the conclusion of each story, the Nyankusi were anxious to move on to the next story/chapter.

At this point, Pastor John utilized contingency. He explained that to hear the next story, the Nyankusi had to first tell the story they had just heard to the rest of their Boma—teaching them.

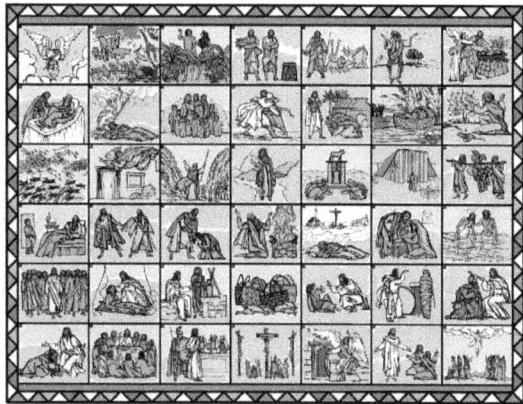

Figure 7.8. The story cloth was used to tell the stories of the Bible to the Maasai.

Brilliant strategy! A win-win-win approach. The Nyankusi fulfill their responsibility of teaching, then they get to hear the next story, and the missionaries get to tell the next story they want to share.

More on Contingency

Some try to put a label of manipulation on contingency. However, that is certainly not the case. Contingency is life. When you come to work, then you get paid. When your daughter finishes her homework, then she gets to go outside and play.

Lastly on contingency—one thing that should never be contingent is love. Love is given because of who you are, not what you do. When love is withheld based on performance, the relationship will be destroyed.

THE STORY
Alabama Sheriff's Boys Ranch

Through the years as I shared my continuous improvement adventures with my father-in-law, David Swanger, he took a keen interest in the principles and applications. At one point in his retired life, he and his wife Skippy served as house parents at the Alabama Sheriff's Boys Ranch in St. Clair County. Here in Pop Swanger's own words is his story:

From Chaos to Harmony on the Boys' Ranch

My name is David B. Swanger, and I am 59 years old and retired. In November 1985, my wife "Skippy" and I found ourselves drawn into the childcare field. We were asked to supervise one of the residential units of the boys' ranch. After a somewhat lengthy interview, we were asked to spend the night in one of the ranch houses to get acquainted with the residents. In the confines of that residence during that span of time we came to a clear understanding of what is meant by the word "chaos." Skippy and I knew that we would not be able to survive in such an atmosphere. The boys were unruly and loud and there were acts of violence and abuse. We began to have second thoughts.

Not long before this, while visiting my grandchildren (and their parents) in Kingsport, Tennessee, it was my good fortune to be able to attend a short seminar on Performance Management. By a happy chance, my grandchildren's father, Russell Justice, was the principle speaker at that session. I heard such things as "Behavior is a function of its consequences," the ABCs of behaviors, NICs and PICs, and other choice behavior-modification tidbits. Let's see, I thought, behavior elicits a re-sponse which (if appropriate) becomes the antecedent that tends to re-duce errors in behavior to zero. I recognized interesting parallels with my work of 35 years in process control.

By happy coincidence, Russell and family came home shortly after our first troubled visit to the ranch. When I asked whether we could design a Perfor-mance Management approach for the situation at the ranch, he was very confident that we could. After much discussion between Russell, Skippy, and me, we were able to identify seven categories of behavior that we want-ed to measure and improve. We designed our scoreboard in such a way that if the behavior or activity was good there was an "x." If not, the infraction or problem was written in that space. Good scores or significant improve-ments would be highlighted in red. Rules were written which explained re-wards and how the team or individual score was used to determine who was eligible for them. A daily score of 6 out of a possible 7 earned the right to watch TV that day during television hours. 42 out of a possible score of 49 for the week earned the right to have a VCR show during the weekend. The high scorers had the privilege of helping the houseparent select the movie to be shown. Weekly average team scores were plotted on a graph. When "best ever" weekly team score was equaled twice or exceeded once in any month there was a special reward such as bowling, skating, or the movies.

When we presented our system to other staff members, they showed gen-uine interest. One individual suggested converting scores to "percentag-es" and we immediately accepted the suggestion. It was a good idea, and we knew that if we asked for the input of other staff members it would enhance our chances of success.

Our confidence at this point led us to accept the position of housepar-ents, and we were on our way! But our kids were skeptical at first. It took a little while for them to realize that we were using a system that would be fair to all.

We soon discovered that occasionally all the boys on the ranch were be-ing loaded on a bus and taken to various places for entertainment and recreation. Everyone received this reward regardless of past behavior, and we told ourselves right then and there that our boys would be rewarded for their right actions—and not for just "being there."

The scoreboard was prepared and put on the bulletin board at night after the boys had gone to bed. It was always interesting to watch the boys assemble at the scoreboard every morning before sitting down for breakfast. They were eager to know how they had done the previous day.

One fine day after a very trying (but satisfying) work detail, the boys and I were all aboard the old "Blue Goose" headed back to the ranch house. Joey Whitney, who was a resident of one of the other houses, was riding along with us and sitting in the cab with Michael and me. Joey spoke up and said, "Oh, I wish I was in the Beeson house." And Michael answered immediately, "Everybody does!" At that time, our Performance Management system had been in operation for about two months and hearing those words pleased me greatly.

During the first 4 months that this system was in operation there was a gradual but distinct improvement in the behavior of all the boys in the house. At one of our staff meetings, we were greatly encouraged by the remarks of a fellow houseparent. He told us that both he and his wife had noticed how much happier the children in our house had become since we assumed our houseparent roles. The numbers, scoreboard, and graphs do not really tell the whole story. The data cannot really convey the feelings of peace, love, and harmony that emerged there. Our system changed us all, and we soon began to think of that ranch house as our happy home.

CASE FILE

A New Pair of Baseball Shoes

Son Matthew enrolled in T-ball for the summer and soon came to me saying that he needed a new pair of baseball shoes. He said, "All the boys have them." After checking around, I found out that not *all* the team had new shoes. (If they did, we would have headed to the store for new shoes.) But, since not everyone had new shoes, I hatched a plan.

Matthew was learning the game, but hitting was not coming easy. So, the contingency was this: When Matthew got his first hit in a game, we would get him new shoes. It took a couple of games, but left-handed Matthew hit one down the third-base line and headed for first with his first hit. Before he even made it to first base, the coach looked up in the stands to make eye contact with me and yelled out, "Looks like a new pair of shoes running down to first base."

You see, Matthew had told the coach (and probably his friends also) about the deal (contingency)—first hit → new shoes.

Now, I should point out how Matthew cared for those shoes. Unlike his old shoes that were often dirty, muddy, and scuffed up, he cleaned the new ones every time any dirt or mud got on them.

CASE FILE
Royal Ambassador Recreation Time

My best friend John and I taught the third- and fourth-grade Royal Ambassadors at our church. (Royal Ambassadors is a boys' organization to teach and do mission actions.) We had worked with other groups before, but this group was by far the most challenging—a force to be reckoned with. After weeks of calling the boys down, threatening to tell their parents about their bad behavior, and doing everything we could think of to achieve calm, we still had mayhem.

We had to come up with something to help us survive. That something was a whole new system. The way the weekly 1-hour sessions were set up, we had 40 minutes for class time followed by 20 minutes of recreation time.

Our genius new system worked like this: The first week of implementation I went to the writing board, drew a box, put the number 20 inside the box (representing our 20 minutes of recreation time), and we started our lesson. In just a couple of minutes, one of the boys broke his pencil and asked the boy next to him to pass him another one from the pencil box. The boy's response was, "Get your own pencil, you dummy. You shouldn't have broken it to start with."

I walked to the writing board without saying a word, marked through the 20 and wrote 19. Before long, another boy decided he was tired of working on the assignment, crumpled up his paper, and threw it at another boy. I walked to the board, marked through the 19 and wrote 18. By now, the boys were paying attention to the board but didn't say anything. The next noticeable event was when one of the boys remembered a scripture verse we had been memorizing and gave it as the answer to a question. Up to the board to change the 18 to 19.

Then one of the boys piped up, "Is that the number of minutes we have for recreation time today?" My answer, "Yes." And "We can add minutes too?" was the next question.

Without really doing a good job of specifying what smart mouthing, roughhousing, or rowdiness would not be permitted and what participation, paying attention, and teamwork would be rewarded, the system was in place and the definition of what was desired was forming very quickly.

As you would imagine, the whole class changed and before long, we had to limit weekly recreation time to 30 minutes and put any extra minutes earned into a "savings account" for a weekend outing.

It's All in the Way You Say It

We had a system around our house: Make your bed up before school; watch TV after doing homework. One day I heard one of our children talking on the phone to a friend and saying, "No, I didn't get to watch that. Mom wouldn't let me because I forgot to make up my bed this morning." Whoa, this system was either designed wrong or explained wrong. Mom didn't keep her from watching TV; she kept herself from watching by not making up her bed.

We then attempted to explain the system and make it clear that if you make your bed before school, then you get to watch TV after homework. It doesn't matter to us if you watch TV or not; it's your choice. I later heard her on the phone with her friend saying, "No, I didn't get to watch it. I forgot to make up my bed this morning."

How to Reinforce?

1. Natural: Job gets easier
2. Social: Primary, essential, what you do or say—bragging
3. Tangible: Optional, symbols of accomplishment

Three issues face us when it comes to *how* to reinforce: the use of natural, social, and tangible reinforcers. All are important and have a place. In combination, they provide the fuel to drive improvement.

Natural Reinforcers

- Come from the behaviors and accomplishments themselves (intrinsic).
- Do not have to be "delivered"–the job gets easier.
- The challenge is making natural reinforcers visible–getting people to see that by improving the process, their job becomes easier or more satisfying. You can accomplish this by discussing "What's in it for me?" (WIIFM) during team meetings, making a list, and posting the list of natural reinforcers (personal benefits in the work area).
- Example: By reducing the damage to bales of filter tow, operators did not have to unpack the bales and manually feed them back into the press, which was a physically demanding task that no one liked.

Social Reinforcers

- Consist of what you do or say (interactions), not what you "get."
- Involve your time and attention.
- Cause discussion of results and supporting behaviors.
- Create a chance to brag.
- Social reinforcement is primary and essential.

Tangible Reinforcers

- Facilitate the delivery of social reinforcement.
- Require out-of-pocket money (but should not be expensive).
- Use symbols of the accomplishment and actions (e.g., 3 Musketeers bar for 3 weeks in a row of on-time delivery).
- Create a story to tell.
- Provide future opportunities to relive the accomplishment by having a "trophy."

- Have no "salvage value." People not associated with the effort and accomplishment would not value it.
- If you question whether it costs too much, it does.
- Tangible reinforcement is secondary and optional.

In reality, the distinction between social and tangible is not cut and dried; there is a continuum between social and tangible. And the best reinforcement is a combination of the two.

Both social and tangible reinforcement create a story to tell.

You have heard the quote, "Sticks and stones may break my bones, but words will never hurt me." That's total nonsense! Remember, words can be life-giving or personally painful. They can destroy or build up. No amount of tangible reinforcement will ever take the place of social reinforcement—what we do with people and what we say to them.

Don't let reinforcement become "give me something." Mix up the celebrations to include things like a dunking tank, putting a plaque on the wall, or a dart throwing contest.

There is no present like your time. There is no better way to reinforce someone than by asking them, "How did you do it?" and then giving them your listening ears.

Reinforcement plans should address four areas:

- daily behaviors (good practices)
- actions for improvement
- milestones and significant events
- reaching goals

There is no present like your time.

Symbolic Tangible Reinforcement Is Cognitive

Your 12-year-old son comes home from school and tells you that he took a brain pill at school today. This immediately gets your full attention, and you ask him to tell you more about this. He explains that every morning, the teacher gives a one-question quiz to see if your brain is awake. If you miss the question, you must take a brain pill.

Wondering what in the world is going on at school with this teacher, you dig into this story some more. You ask him to tell you more about this brain pill. He says, "Oh Dad, it's a jellybean." R e l i e f.

But something seems not right here, and you say to him that it seems backward to you, that you get to eat a jellybean if you miss the question. Your son looks up at you with a disgusted look and says, "Dad, you don't *get* to eat a jellybean. You *have* to eat a jellybean."

It's not how the jellybean tastes in your mouth; it's what the jellybean stands for. It's not what the reinforcement *is*; it's what it *stands for*. Reinforcement takes place not in your mouth, but between your ears.

> *It's not how the jellybean tastes;*
> *it's what the jellybean stands for.*

From the Case Files—Social Reinforcement

These social reinforcement examples from the case files can be used as is or to stimulate new ideas for you.

CASE FILE

Record-Breaking Ceremony

When a record (best ever) is broken for an important measure of performance, buy an old record (remember those disks with a hole in the middle?) from the antique store. (You will need one of the older shellac records that are brittle and will break easily. Don't try to break the vinyls.) Print a new label for the record with the old best ever on it. Have a celebration where you discuss the new record and a team member uses a hammer to smash the old record. Give a piece to each person to take back to their workplace or home as a souvenir of the accomplishment. Write the new record level on another label, put it on another antique record, and display it in a prominent place.

📑 CASE FILE

Parking Lot & Gate

Meet people in the parking lot, entry door, or at the gate as they arrive for work. Have a handout detailing a significant accomplishment being celebrated, such as startup of a new operation 6 weeks ahead of schedule. Ask the employees how they have been involved in the initiative.

📑 CASE FILE

Doorbell for the Cook

This is for restaurants where the check is paid at the cash register. Place a doorbell on the wall and a sign that says, "To praise the cook, ring the bell." When the doorbell rings in the kitchen, the staff there applaud the cook.

📑 CASE FILE

Take a Whack!

When you beat a competitor to market with a product innovation, "take a whack" at the competitor's product by placing it on a block of wood or stump and smashing it with a sledgehammer.

📑 CASE FILE

Torpedoing the Competition

Each time the lab completes a major product innovation that will do damage to their competitor, loudspeakers in the lab ring out with sonar beeps and "dive, dive." John and his teammates know to go to the conference room to hear about the latest innovation. There they find the Lab Chief, dressed in his naval uniform, ready to tell the story and place another damage sticker on the lab's mockup torpedo.

📑 CASE FILE

New Year's Party in November

When sales for the year cross the annual goal on November 19, management throws a "New Year's Eve" party. When an article about the party is published in the local newspaper, friends and neighbors tease company employees about not knowing when New Year's Eve occurs. They respond that it was November 19 for them this year because they had completed a full year's worth of work by then, and the remaining days of the year (42 days) are bonuses.

📑 CASE FILE

We're in the Gravy Now!

When global sales of $8.3 million per day far exceeded the best year ever ($7.1 million per day) by December 5, biscuits and gravy were served to thousands of employees in over 20 plants and offices around the globe to celebrate being "in the gravy now." The rest of the year was gravy (extra).

The biscuits and gravy were served up by management in each location, with congratulations as employees came through the serving line. Graphs showing the factors driving this increase filled the walls. (Note: In some countries it was necessary to explain what gravy was and that a biscuit was not a cookie.)

📑 CASE FILE

9-10-11 Award

This award was given to any sales representative who increased sales in their district by 10% or more, for nine of their top 10 customers, by the end of the 11th month. This would mean 10% more sales for the year with one whole month still to go. Winners of the 9-10-11 Award receive one round of golf on any course of their choice in the U.S. In addition, their district manager has to caddy for them and is not allowed to play.

📑 CASE FILE

Million Dollars Celebration

When the Plant Transportation Department reached $1 million of savings in laundry, truck maintenance, and machine tools services, they had the credit union bring over $1 million in $20 bills (and a security guard, of course) as the centerpiece for the celebration. Members of the department got to have their photo taken with the stack of money, touch it, and smell it. Talk about a story to tell back at home that night! (And, with the company profit-sharing plan, the members received some of those dollars from their hard work back in their pockets.)

📝 CASE FILE

Half a Celebration

When the office was halfway to reaching the goal for getting orders from new customers, they celebrated with half pizzas. They were ordered from the pizza store to not be cut in half, but to be baked as a half pizza with crust across the middle.

When the team was halfway to eliminating the excess-ink inventory, they all stopped work a half hour early and went halfway home to the local tavern for half a sub.

📝 CASE FILE

Microchip Chips

Production of microchips in the high-tech engineering manufacturing area was hitting the standards and at design capacity. At the same time, demand was still rising. So, the obvious solution was to build more capacity. But with some persuasion, it was agreed to try the discretionary effort approach, challenging the workforce to use their experience and creativity to find ways to redefine capacity through innovative ideas and the reduction of waste and downtime.

Given a production goal for the coming year, the workforce was challenged to see how many days *before* the end of the year (December 31) they could reach the annual plan. It was agreed that the remaining days, after the annual plan was met, would be divided in half and used to (1) do some research and study on evolving microchip technology (which was something the engineers loved to do but never had time for) and (2) produce more chips leading to more sales, profits, and profit sharing.

So when the annual production plan was reached on December 4, that left 27 extra days. Taking into account the 2-day Christmas holiday, that meant 13 days for research and 12 days for more production. Even more importantly, it meant a new, reengineered capacity based on the improvements that had been made without spending the capital on an expansion.

As a reminder of the challenge and the rewards, an arrangement was made with the local potato chip manufacturer to produce individual sized "microchip chips" with instructions on the back of the package for how to expand microchip capacity. The bags of microchip chips were made available throughout the production, shipping, and office areas.

Clearly the opportunities for social/symbolic reinforcement can be found throughout our everyday life—like the two examples below.

CASE FILE
The Handshake

On *The Great American Baking Show*, the highest honor is to get the handshake from host Paul Hollywood. When he gives you the handshake, everyone knows you have just created a bell ringer. On one program he even gave two handshakes, and everyone was shocked.

THE STORY
11th-Grade American History & the Mayflower Seafood Restaurant

In the state of Tennessee, every 11th-grade student is required to take American history. More often than hoped, some of the students in the class fail and have to take it again. The coach did not want that to happen in his class. In fact, he wanted every student to make at least a grade of B. He wanted to have an all-star class.

He told his class about his goal and suggested that if everyone in the class made an A or B, he would take them out for lunch at the local hamburger joint at the end of the semester. The next day, some of the students came back with a proposal: They would go to the Mayflower Seafood Restaurant and have Coach dress up as a Pilgrim. He agreed to the plan. If everyone in the class made an A or B, he would take them all out to the Mayflower dressed as a Pilgrim.

A few days later, Coach invited the home education teacher to come into the class and measure him for the Pilgrim outfit that he would wear. Soon after that, he began to hear students talking about meeting at Shoney's before school to study together. He heard them asking each other if they had read the assignment and if they needed any help getting ready for the exam.

Midway through the semester, Coach came into class wearing a Pilgrim suit and made the announcement, "Here it is. But no one, except you, will ever see me wearing this unless all of you make A's or B's."

At this point, I think he had them. He had made a deal—a contingency. If you ... then I ... . To cash in on their end of the contingency, they had to work together as a team. Each student had to not only make an A or B for themselves, but they had to make sure everyone else did the same. He had eliminated the competition for the A's in the class and created cooperation instead.

American history, Mayflower Seafood Restaurant, a Coach Pilgrim—a recipe for teamwork, learning, and a celebration.

For more social reinforcement examples, see Appendix C: More Social Reinforcement Examples.

From the Case Files—Tangible Reinforcement

These tangible reinforcement examples from the case files can be used as is or to stimulate new ideas for you.

As stated earlier, there is a continuum from social to tangible reinforcement. So, perhaps for these case-file examples, we should say they are "more" tangible.

📑 CASE FILE

Cracker Jacks™

Unfortunately, the box does not say it anymore, but for many years the quote on the box said, "When you're really good, they call you Cracker Jack." I have used the presentation of a box of Cracker Jacks along with some bragging as one of my favorite reinforcers. (You can print your own labels with the slogan and stick it on the box. You can also find and show a video clip of the old commercial about being a "Cracker Jack.")

CASE FILE

Popcorn–Things Are Popping Around Here!

Walking into the Marketing Building one day, I was met with the aroma of popcorn. I asked what was happening, and the receptionist told me to go to the third floor. There I found the VP of plastic sales, standing by a popcorn machine. When I asked what was going on, he explained that "things were popping around here," and worldwide plastic sales had just set a new record. He said that he was just creating the aroma of success.

CASE FILE

Permanent Fix Club

This club meets once a month to hear stories of permanent fix solutions for nagging problems. New members are inducted into the club. Tools for fixing things like a wrench, hammer, saw, glue gun, and stapler are presented as symbols of the achievements. The purpose of the club is twofold: (1) Recognize those who have developed permanent fixes, and (2) replicate those ideas as members hear the stories and take the ideas back to their work area for implementation where applicable.

THE STORY

Cherry Mash

An associate from our sister operations in New York developed a keen interest in the methods of ACI. At the same time, he was experiencing rapid aging as the result of a disease. Deciding not to retire, he transitioned from his assignment as a supervisor to an ACI facilitator, saying he enjoyed it so much he wanted to spend some of his remaining days helping others use these methods. On a trip to our Tennessee operations, he presented me with a Cherry Mash (a giant chocolate-covered cherry) as a gift for introducing him to ACI. He told me that he received two of these Cherry Mashes each year from his mother who lived in Joplin, Missouri, where they were made. He received one for Christmas and one for his

birthday. I treasured his gift for years until it disintegrated, but I kept a photo for memory's sake.

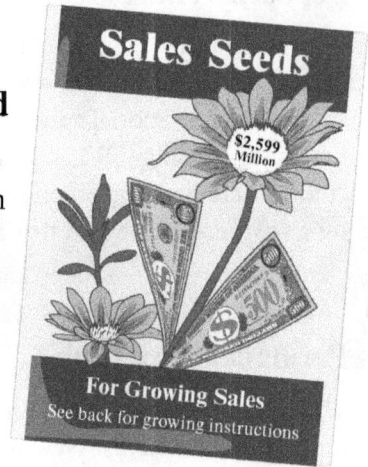

📋 CASE FILE
Sales-Growth Seeds–Customized Tangible Reinforcement

To recognize and celebrate reaching an all-time record for sales revenue of $2,599 million far ahead of plan (November 7), as part of the celebration, employees were given packs of "sales seeds."

On the back were the "instructions for growing sales":

- Start by knowing customer needs.
- Add product uniformity, process stability, defect prevention, waste reduction, yield improvements, and reliability increases.
- Accelerate innovation.
- Improve customer interactions (sales calls, order taking, delivery, service).
- Then harvest the sales-growth revenue, reinvest in the future, and reinforce the behaviors while celebrating the results.

📋 CASE FILE
Certs & Life Savers®

Certs were given for being "certified." We made a new label that specified the "ingredients" for being certified. Life Savers were given to someone for "saving our life." You can print a label to stick on the Life Savers that tells the story of the rescue.

Note: If people eat the candy and throw away the wrapper, the event probably did not have the maximum impact. We want the treats (and more importantly the wrappers) to be taken back to the work area or home like a trophy.

CASE FILE

Free Lunch

When employees are working on a project to meet the production schedule, don't charge them for meals in the cafeteria as long as production is on or ahead of schedule.

For more examples, see Appendix D: More Tangible Reinforcement Examples.

THE STORY

That Won't Work Here: Safety in Mexico

The project was to build a multimillion-pound PET bottle-polymer facility in Mexico. The facility was a greenfield plant, a from-the-ground-up construction project. Based on the Mexican labor consultant's description of the general safety in the Mexican construction industry, there was concern about safety for the project.

Working with the construction team and the labor unions, Total Quality Management (TQM) consultant Jim set out to formulate a safety plan with the ultimate goal of no lost-time accidents. Jim was told by the local labor leader, Juan, that symbolic tangible or social reinforcers would not work with Mexican craftsmen, adding that the only thing that worked was more pesos. So, the challenge was to discover a means for recognizing a hard-working and proud labor force—with different needs than those of their American counterparts—and to do so in a way that fulfilled those needs for recognition while showing respect for the culture and preserving dignity.

With determination and creativity to find meaningful reinforcers, reinforcement looked like this as the project progressed:

- At 125,000 work hours without a lost-time accident, and at the suggestion of Juan, each worker on the project received a package of grocery staples and supplies (such as corn meal, flour, beans, and sugar) and an attaboy for working safely. It was rewarding to observe the acceptance of the packages.

- After canvassing the workforce regarding a desirable reinforcer, at 250,000 work hours with no lost-time accidents, each worker received a long-sleeved, chambray-type work shirt.

- Realizing that the crews considered practical items most reinforcing and after observing daily work, it was noticed that the workers were sometimes filling their hard hats with water (from the glass jugs that were brought in each day) for drinking and washing up at the end of the day. Insulated Coleman® thermos jugs with the project logo were ordered for everyone and handed out with a word of thanks for continued safe work.

- To give workers a place to secure personal items while on the job site, fanny packs were distributed to the teams when a subgoal was met.

- When the millionth work hour with no lost-time accidents was reached before the end of a year, managers attended a celebration where they congratulated the workforce and personally distributed hats and shirts monogrammed in Spanish. By coincidence, the celebration fell on the Mexican holiday of Día de la Santa Cruz—a day that honors the stonemasons of Mexico.

The new plant began operations on schedule. Far from fulfilling the prediction of multiple safety problems, the workforce instead set an impressive record of over 3 million work hours with only one lost-time accident. Fortunately, that one accident, which occurred after a full year of construction and over 1 million work hours, was a minor one that kept the injured worker off the job for only 1 day.

At the conclusion of the project, the contractors told us that a new standard and benchmark had been set—not only for their company, but for the entire country of Mexico.

For more about this story see:

- "That Won't Work Here": Positive Reinforcement in Mexico—PM Magazine, Vol 15, #2, page 3. © Aubrey Daniels International. Used with permission. https://www.aubreydaniels.com/media -center/archive/performance-management-magazine-spring-1997 -vol-15-no-2

The Best Reinforcement

You can see clearly from these case-file examples that the best reinforcement is a combination of tangible reinforcement (symbols that represent the improvement and create a story to tell) and social reinforcement (talking about what we did). The tangible reinforcement creates a reason, an occasion to come together. The real reinforcement takes place through the words that explain the accomplishment, why it's important, and how it was done.

✎ TIPS

Tips for Selecting Reinforcers

- Keep natural reinforcement visible. (How will this improvement make our job easier and improve our product or service?)

- Use praise and recognition as the primary reinforcers (words and actions).

- Use facilitating tangibles (such as food) to provide a relaxed setting. (It's easier to talk about what we have accomplished and how we did it with a cup of coffee in our hands.)

- Use symbolic tangibles to create a memory: Life Savers, "you nailed it" nails, steak cookout for putting a stake in the ground, 7UP for uptime.

- The cost of recognition is never a barrier. Exceptional reinforcement can be generated with creativity, not by spending big money. When I hear supervisors and managers talk about "incentivizing" a behavior, I cringe. They are practicing lazy management, not doing their job, and are about to waste a lot of money.

- All tangible reinforcement should be symbolic. Gift certificates are not good reinforcers. They represent lazy management.

- Use a variety of reinforcers to prevent satiation. (When the same social or tangible reinforcement no longer satisfies, it becomes like overdosing on chocolate.)

- You cannot "buy" reinforcement. You must create it with the symbolism you invest in it.

- Do not escalate tangible reinforcement. Fast-food hamburger, gourmet hamburger, hamburger steak, sirloin steak: Do not get caught in this trap. Mix it up.
- The goal is to have people talking about what we did, why it is important, and how we did it—not what we got.
- Remember that *how* reinforcement is given is more important than *what* reinforcement is given. Consider the amount of time put into identifying the what. Even much more time must be put into the how. What will we say? Who will say it? Who will be there? How will we bring out the actions that got us here and then recognize them?

Planning Reinforcement

- Creates more opportunities by setting up occasions for reinforcement.
- Sensitizes you to look for opportunities.
- Increases creativity by allowing time to think.
- Serves as a reminder (trigger) of what to reinforce.
- Helps to create a balance of recognizing results and celebrating the behaviors leading to the result.
- Stresses a series of reinforcing events.

Avoid When Reinforcing

- Assuming the average person does not *warrant* recognition.
- Assuming the average person does not *want* recognition.
- Waiting too long to reinforce. (Do it when sweat is still on the brow.)
- Coupling reinforcement with criticism or asking for more ("You did great, but ... ").
- Recognizing only after the results are in. (Recognize milestones along the way.)
- Assuming "one size fits all" when choosing methods for reinforcement.

Managing Reinforcement

- You can *do* a lot of reinforcement yourself; there are lots of opportunities.

- You can *cause* a lot of reinforcement by the design of the system that creates opportunities (even in the middle of the night while you are sleeping).

- Your *example* can be a model for others. (As author John Maxwell says, "Your walk talks louder than your talk talks" [Maxwell, 2023].)

"The systems in the excellent companies are not only designed to produce lots of winners; they are constructed to celebrate the winning once it occurs."

—Peters & Waterman, 1982, p. 58

Reinforcement and Themes

There is a level of reinforcement you can't achieve unless you use themes. The rate at which things improve is a function of the amount of reinforcement received. Themes multiply the reinforcement many times over.

Some Theme Examples

- An aerospace company's "on-time program performance"—on a safari to Busch Gardens. Each goal achievement got everyone closer to a day at Busch Gardens.

- "Climbing the pyramid of customer success" for Latin American sales.

- Equipment uptime—"on track" (with NASCAR theme).

- Superheroes comic books for providing "super service."

- "Wrapping up success" for improving gift wrap production.

Gemba Walk

In his book *Gemba Kaizen* (2012), Masaaki Imai tells a story about Taiichi Ohno, the creative inventor of lean manufacturing. When Ohno would notice a leader out of touch with the gemba, he would take the leader to the plant floor or office, draw a circle on the floor, and have the leader stand there until he began noticing opportunities for improvement (kaizen).

What if you had to stand in the circle until you saw something good, something to reinforce?

Go to the gemba (the actual place—shop floor, office, kitchen). See for yourself.

- Treasure hunt to find things being done right (using the right tool, using checklists, using correct procedures, having a clean and organized area, and keeping up-to-date data records).
- Tag and tell an individual or group by filling out a tag (special design and color) telling them what you saw that you liked, who did it if you know, your name, and the date.
- Talk to a person there and give them the tag. Ask them about the work, critical steps, and impact on the product or service.
- If no one is there, leave the tag. Place it on a scoreboard, wire it to a machine, tape it to a clip board, leave it on a desk.
- Tell the area supervisor about your tags.
- Talk about tags in the next team meeting and incorporate good practices into normal operations.

Two Obligations

Remember the two obligations. Anytime a leader asks the team or an individual to do something, the leader has two obligations:

- Observe when it is done.
- Recognize when it is done.

Through the years, I have encountered some pushback on this from leaders who say they don't have time to do that. I say that if you don't have time to observe and recognize, then don't ask for it to be done. In the long run, this will likely reduce the number of things you ask people to do—and that will be a good thing, helping the organization to focus.

Dancing in the End Zone

Consistently doing the right thing, establishing a best ever, or reaching a goal calls for stopping what we are doing for at least a moment to talk about the success and enjoy it. Get reenergized before we get back to work. It sounds simple enough, but often we just don't do it.

For me, it's like dancing in the end zone after a touchdown is scored. It's natural. It's just pausing for a moment to enjoy what we've done. Then, it's back on the field to play defense.

Think back to the best day you ever had at work. What happened that day? I suspect it involved some recognition or reinforcement. Recall that day and seek to create a similar situation with your celebrations.

Celebrating Results

We have reached our goal, and it's time to celebrate. We've assembled everyone who contributed to the success, along with the customer who is benefiting from the improvement. But what are the vital elements for celebrating an accomplishment or result? There are four critical steps for doing it the right way:

- Talk about *what* we have done (show the graph).
- Talk about *why* it is important (to customers, company, and workers).
- Talk about *how* we did it (identify behaviors and learn).
- Enjoy the success (with social events and symbolic tangibles).

Let's dig into each of these steps and talk about the details and good practices.

What We Did

- Recall how this focus for improvement was selected.

- Show the data (graph, chart, illustration). We're here because something improved. No graph → no celebration.
- Don't just show the graph. Leave it posted for all to see what got us here.
- Show the baseline, the goal, and the journey. See that something special has occurred.

Why It Is Important

- Talk about why the improvement is important to the customer and the organization.
  - ○ The best person to talk about why this is important is the customer who benefited from the improvement.
  - ○ If you can't find a customer to testify about the improvement, then you should question having a celebration.
- Talk about WIIFM: "What's in it for me?" How will this improvement make my life better; make my job easier; and provide benefits to me, the worker?
- Make natural consequences visible. Verbalize them and show a list.

How We Did It

- Get people talking about what was done to bring about this improvement.
  - ○ Identify the behaviors leading to the results.
  - ○ In the locker room or in postgame interviews after a win, what do people talk about? They do not just repeat the score over and over. They talk about what they did, what was done, the big plays.
- Bragging rights come from knowing how we did it. If you don't think people will speak up, seed the conversation by asking some people ahead of the celebration to be prepared to share.
- As people share, post what they say for all to see, not just hear. Use whiteboards, flip charts, and projectors.
- Emphasize data, facts, root causes identified, problem-solving, and innovations.

Celebrate the Accomplishment–Enjoy the Success

- It would be OK to meet, talk about what we did, why it is important, and how we did it and then get back to work. But that wouldn't be enough. Why not take a few more minutes to just enjoy the accomplishment with some refreshments or tangible presentations?

- Create a story to tell through a fun event.

- Use symbolic tangibles to create memories and make them last.

Cautions

⚠ CAUTION

4:1 Rule–Spend four times as much time planning what you are going to say as what you are going to serve. I once received a phone call from a team that was planning a celebration, and they were having a hard time deciding whether they should have two kinds of pizza (cheese and pepperoni) or just have cheese and keep it simpler. I must admit, it was not kind of me, but I jumped on them, telling them they were focused on the wrong topic. Who cares about the pizza? They should worry more about the symbolism of the pizza than which toppings to get, and be spending their time planning what to say during the celebration. This is the 4:1 rule (Madsen & Madsen, 1974, as cited in Daniels & Rosen, 1989). (Note: They really shouldn't be eating pizza unless it is a symbol of what they have accomplished; like "Bull's-Eye Pizza" for hitting an on-time goal for delivery.)

⚠ CAUTION

MEBW–This acronym stands for a distorted version of the four steps for celebrating. MEBW is Meet, Eat, Burp, and go back to Work. It has been said that lots of reinforcement is given, but little is received. If your four steps of celebrating are (1) meet, (2) eat, (3) burp, and (4) go back to work–that is not a celebration; it is just eating together. I once heard two people leaving a celebration, and one said to the other, "What was that all about?" The second person replied, "I don't know, but these are the best chocolate chip cookies I have ever eaten." If you asked the supervisor, they would probably say they had a "celebration." I would not agree; it is only a celebration if the real four steps are followed:

- Talk about *what* we have done.
- Talk about *why* it is important.
- Talk about *how* we did it.
- Enjoy the success.

Lots of reinforcement is given, but little received.

⚠ CAUTION

Gift Certificates—No! Lame idea. Not personal. No story to tell. Lazy reinforcement. The way you show that the accomplishment is important to you is by investing some time in creating the reinforcement. Reinforcement cannot be bought; it must be created.

From the Case Files—Celebrations

📑 CASE FILE
One Scoop or Two?

There's nothing like a banana split to cool off on a hot afternoon. Engineering Division directors and managers served over 40 people who had contributed to a successful major shutdown and startup of the coal gasification unit. The event was held on the lawn of the headquarters building to make it high profile and to bring attention to the team that had significantly shortened the targeted shutdown time, getting the operation back up and running in record time.

📑 CASE FILE
World's Largest Car Wash

To link into the overall company pinpoint of sales revenue growth, the Plant Maintenance Division selected "on track"—keeping critical pieces of equipment on track by reducing equipment failures and reducing repair times. A goal of a 1% increase in the equipment reliability index (ERI) over the next 52 weeks was set. With this challenge came a promise to celebrate achieving the goal with a car wash for over 1,000 maintenance employees: a car wash to be given by their supervisors and company management.

At one of the plant gates at shift change time one day, I heard a division employee who was heading home talking to another employee just coming to work. He told him, "We've been having trouble with #6 all day. Don't let it go down tonight. I'm looking forward to having Bob [the company president] wash my car."

In July, the 52-week goal of at or above 1% in the ERI was met. (Nice timing for a car wash.) In fact, the ERI for the last 52 weeks had increased by 2.6%, resulting in a value of $20 million in increased earnings from the increased uptime, production, and sales!

The day-long celebration started at 7:30 a.m. and went until 4:30 p.m. (see Figure 7.9). There were refreshments, entertainment, and door prizes. A NASCAR race team was on-site so people could sit in a race car, take photos with, and talk to the pit crew and driver. Over 1,000 cars were washed, with some employees going back home to get a second car to be washed.

Figure 7.9. The promised car wash was a grand success.

📑 CASE FILE

Make International Business Easy (MIBE)

To celebrate the staggering success in the aim to make international business easy (MIBE), improving customer satisfaction and growing sales revenue, an MIBE Festival was held at company headquarters. Teams from around the globe gathered for a day of celebration. The festival included international flags, music, food, and attire.

Storyboards were displayed to share project results.

For more about this story see:

- MIBE – Make International Business Easy—PM Magazine, Vol 8, #4, page 19. Aubrey Daniels International. Used with permission. https://www.aubreydaniels.com/media-center/archive/performance-management-magazine-fall-winter-1990-vol-8-no-4

- Making International Business Easy (MIBE)—PM Magazine, Vol 9, #4, page 10. © Aubrey Daniels International. Used with permission. https://www.aubreydaniels.com/media-center/archive/performance-management-magazine-fall-winter-1991-vol-9-no-4

📑 CASE FILE

Mountain-Climbing Axe—The Sherpa Award

They showed up at my cubicle, the executive team. The company CEO said they had come to see me to express their thanks for being their sherpa as we had climbed the mountain of sales revenue growth from $7.1 million per day to $8.3 million. They presented me with a mountain-climbing axe.

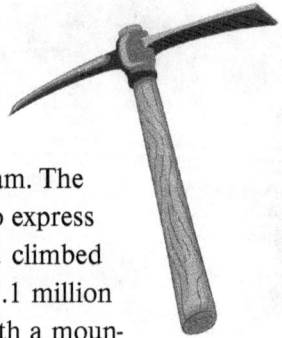

Before heading back to the headquarters building, they visited and we talked for a few minutes, recalling the struggles, challenges, and successes along the journey up the mountain. When they left, the office was abuzz, as no one recalled a visit, ever, to Management Engineering by the entire leadership team.

Did I feel appreciated? Was I proud? You better stand back away from me because I might explode with joy. The leadership team! In my office! I had been to their offices many times, but here they were making the journey across the plant! Obviously, they had spent time planning a symbolic reinforcer. They took the time, not just to say "thank you" but to travel to where I was, to me. They showed me some love—and some say that love is spelled T-I-M-E.

THE STORY

Let'er Fly

The month marked the special event of daughter Lauren's graduation from the University of Tennessee (UT; BS in Broadcasting). With graduation came the challenge to celebrate the occasion and pay tribute to Lauren in a way that exhibited our love, pride, and hopes for her. Having spent many years teaching about and designing celebrations, I thought to myself, "Now it's time for the proof in the pudding—the time for me to create a real bell ringer celebration." Debby and I had some good plans but still needed the ringer idea—that one special activity or gift. We asked for input and ideas from our trusted family and friends, but the week of graduation was approaching, and we were still searching. Then the idea came—graduation as an arrow being removed from the quiver and shot out into the world to make an impact.

In our graduation celebration with family gathered at a cabin in the Smoky Mountains, we presented the arrow saying, "We have here an arrow—a real arrow, specially chosen for you. I did not know before I began the search for this arrow that the shaft comes without an arrowhead. This arrowhead was selected with great care and attached using a torch. It is an arrowhead that will make serious impact to the things it comes in contact with—to the things it impacts. This arrow is designed to be appropriately guided by white and orange fins—symbolic of the training you have received from UT. Lauren, this arrow represents you. Your life.

"It has been said, and rightly so, that arrows are not meant to stay in the quiver. They are fashioned to be shot out into the world to make a difference, and we are very excited about the difference you surely will make.

"Your graduation today marks your launch. You are ready for flight. It is a bit scary to us, but even more exciting. We wait with great anticipation to see just how you will impact your world. We know you will have a positive impact. We anticipate a significant impact where you work in the media field. And later (when the time is right), we can only begin to imagine how you will fulfill your roles in life with great excellence.

"Lauren Brooke Justice—'victorious one'—FLY!!!"

✎ TIPS
When Planning Celebrations
- Include all members of the workplace team.
- Include others who contributed to the accomplishment.
- Plan the celebration for as soon as possible after the accomplishment.
- Conduct the celebration in the work area.
- Invite customers and others who have benefited from the improvement.
- Make the reinforcement from you, not from "the company."
- Ask higher management to attend.

Celebration Scorecard

Figure 7.10 shows a scorecard for evaluating your celebration. Use this after each celebration to evaluate how well you did and to make improvements.

Reinforcement Summary
- Celebrate results and reinforce behaviors leading to the result.
- Use a combination of natural, social, and tangible reinforcement.
- Social reinforcement is essential. No amount of tangible will ever substitute.
- Use tangibles to create a story to tell and memories.
- Reinforcement is earned, not "given" (contingency).
- Shaping—reinforcing small improvements—accelerates improvement.

- Any system that has a limit to the number of winners is a losing system.

- Design systems for supervisor, peer, and customer reinforcement.

- Having a theme increases opportunities for reinforcement.

- The purpose of a goal is to provide an opportunity for reinforcement.

- How reinforcement is given is more important than what reinforcement is given.

Celebration Scorecard

1.	The measure/graph for performance shown, discussed, and posted. (35 points)	
2.	Value for company, customer, and unit discussed. (25 points)	
3.	Behaviors leading to the accomplishment discussed. (30 points)	
4.	Celebration took place within 1 week of accomplishment. (5 points)	
5.	Customer attended. (10 points)	
6.	Customer spoke. (5 points)	
7.	Celebration took place in the work area. (10 points)	
8.	Future goals and expectations discussed. (*Subtract* 20 points)	
	Total score	

Note: Partial credit is given for each component based on achievement.

If the total score is:
- 0–69: You wasted your time and money.
- 70–79: You know you can do better.
- 80–89: You are nearly there.
- 90–99: Very good.
- 100–120: Fantastic–congratulations on a great celebration.

Figure 7.10. The Celebration Scorecard will help you evaluate and improve your celebrations.

Celebration Summary

Good celebrations follow four steps:

- What we did.
- Why it is important.
- How we did it.
- Enjoy the success.

Ready-Set-Go—Reinforce Behaviors & Celebrate Results Checklist

The leadership team must own the reinforcement plan, especially those parts that impact the entire organization. That means they should be actively involved in planning the celebrations and should participate when they are conducted. Leadership should provide the guidelines for all reinforcement and participate when appropriate.

- ☐ Start your reinforcement plan by identifying key behaviors, best-ever levels, best-ever days/weeks/months in a row, and milestones to be recognized.
- ☐ Brainstorm ideas for recognition (social and symbolic tangibles).
- ☐ Make natural reinforcement visible.
- ☐ Pair the brainstormed ideas with the appropriate milestones and events to be recognized and celebrated to create a reinforcement plan.
- ☐ Plan celebrations by following the tips for planning.
- ☐ Execute the plan by following the four critical steps of a celebration.
- ☐ Evaluate your celebration using the Celebration Scorecard (Figure 7.10) and improve.

CHAPTER 8

Putting It All Together

We have described each of the seven elements of "what leaders do"–the elements that lead to Accelerated Continuous Improvement (ACI), organization excellence, and a competitive advantage.

While you have seen pieces of each case study already in earlier chapters, let's put it all together by looking at some case studies that illustrate all the elements.

CASE STUDY

On Track–Equipment Utilization

Situation	With 1,140 Maintenance Division associates caring for thousands of pieces of equipment, the challenge was to keep the equipment up and running without interruption.
Focus/pinpoint	Equipment Utilization: Using a NASCAR theme of "on track," the focus was to keep the most critical pieces of equipment out on the track and out of the pits.
Kickoff	• The green flag was dropped, and engines roared when kickoff meetings were held with all employees. • The division head explained the responsibility of the division to maintain critical assets in order to increase output, sales, and profits. He also explained that the division needed a common focus and direction, and equipment reliability provided this. • Department superintendents explained their link-ins, goals, and plans to reach them. • Each crew team was asked to identify the 10 most critical pieces of equipment in their area and begin by focusing there.

Translate & link in	• Initially, 24 link-in projects were selected, and teams formed for each. • After the kickoff meetings, associates were asking, "When will my team get involved in the equipment utilization initiative?" All teams were encouraged to utilize the four-step improvement process in their area.
Manage-ment action	• A 3-day, off-site workshop was conducted for the leadership team to draft the ACI plan. • An 8-hour seminar with performance management expert Aubrey Daniels was conducted for general supervisors and department supervisors. • 2-day workshops were held for each department team to learn the principles and concepts and to work on their plan for improving performance in their area. • A team formed with members from the Maintenance, Purchasing, and Stores Departments to improve the quality of spare parts. (As quality raw materials are necessary for quality product, quality spare parts are necessary for quality repairs.) • Improved failure analysis tools and processes were introduced, including attention to vibration, lubrication, and operating conditions. • Partnerships with Operations formed to identify ways to work together in preventing equipment failure and to reduce repair times.
Improve process	• 111 teams met weekly and made significant improvements in their work areas. • Partnerships were formed with Engineering to design out equipment failures.

Measure & feedback	• With the focus on equipment reliability or uptime, there were many possible ways to measure progress and results. For simplicity, visibility, and being rally-able, the measure chosen was the number of critical pieces of equipment that failed during the day. Of course, downtime on some of these pieces of equipment was more costly and the equipment was down for different lengths of time. However, the simple count of number of pieces of equipment going down during the day became the overall measure to be displayed to the organization. Many of the other measures were used in team meetings for problem-solving.
	• A creative way to clearly show status and progress was the use of a Hot Wheels racetrack and cars. The racetrack was mounted on a 4' x 8' sheet of plywood, with the outer track and the pits. A numbered car representing each of the critical pieces of equipment was placed on the track. If that piece of equipment went down any time during the day, it was moved to the pits.
	• At the end of the day, the number of cars that went to the pits was counted and added to the weekly total, which was plotted on the graph.

Reinforce & celebrate	• Teams received a racecar scoreboard by:
	  1. having measures in place
	  2. identifying their 10 most critical pieces of equipment
	  3. establishing preventative maintenance plans for pieces of critical equipment

	• Hard-hat stickers were awarded for "tackles" of problems—identifying the root causes and finding ways to prevent their reoccurrence.
	• Fast food was served for "cars" (equipment) that were running at record-high "speeds" (rates).
	• An overall goal of a 1% increase in uptime for 52 weeks was established. If the goal was reached, a car wash was planned where the company president, division heads, and department heads would wash the cars of the over 1,000 associates.
Results	The actual result was a 2.6% increase in uptime, with a value of over $20 million in additional earnings. The car wash—with sponges, hoses, buckets, and race cars—was a day never to be forgotten. (See more details of the World's Largest Car Wash in Chapter 7.)

📁 CASE STUDY
Railcar Turnaround

Situation	• The railcar fleet is an important and significant asset. The fleet is valued at approximately $500 million, and the current turnaround time is 39 days. For each day that this can be reduced, $28 million can be freed up. Even more important, orders can be missed or delayed because of a shortage of railcars.
	• Many factors affect railcar turnaround time, including the way railcars are cleaned, scheduled, and loaded; plus the transportation of cars; and the scheduling, usage, and return by customers.
	• For the purpose of this project, the focus was on the Railcar Hold Time by the customers.
	• These hopper cars are loaded with PET pellets shipped to bottlers to be blown into plastic bottles for soft drinks and water.
	• With a railroad hopper car costing around $100,000, it is expensive for them to sit idle for several days or weeks waiting to be unloaded.
	• The agreement with bottlers (and the industry standard) was for them to be able to hold the cars for up to 14 days, but that agreement was not being adhered to.
Note	**This is a situation where the performers are not part of the organization—they are the customers.** This illustrates that this approach to improvement is not limited to any set of boundaries.
Focus/ pinpoint	Railcar Hold Time at Customer Locations

Kickoff	• Initial kickoff was conducted with a "pilot" group of customers. This kickoff was conducted in two rounds: first with company management, and then, with their permission, to Production Planning, Purchasing, and material handlers at their site.
	• Emphasized the importance of railcar turnaround for cost control (a 1-day reduction in hold time for the fleet would be a savings of $250,000) and for ensuring reliability of supply (especially in short-term, tight situations like the present).
	• Data for each customer was collected and shared, showing the distribution of days held and the averages.
	• Kickoffs were also held with Traffic Control, Production Planning, Materials Handling, and Marketing–and with the railroads.
Translate & link in	Linking-in took place at customer locations, as each customer was asked to take on this improvement project as a joint effort.
Management action	• Managers held meetings with customers to ask for help.
	• A steering team was appointed to manage feedback and resources.
	• A seminar was held for customers on just-in-time inventory management.
Improve process	Causes of increased hold times identified: • Changes in customer production schedule. • Difficulty in spotting the railcar. • No silo space for storage. • Customer's customer has untypical order pattern. • Startup problems. • Long transit times. • Off-spec product in railcars. • Early shipment to accommodate long run times. • Covering last-minute acceptance of quotes by customer's customer. • Less-used formulas.

Measure & feedback	• Measure: average days hold time at each customer.
	• Reported on a quarterly score card. Showing the number of railcars released back that quarter, the distribution of hold times, average days of hold time, and the number of cars released in 14 days or less. (Notice the emphasis on the number "right," not on the number late.) See the scorecard in Figure 8.1.
	• Percentage of cars returned in 14 days or less for fleet.
	• Baseline average: 27 days for the fleet.
	• First milestone goal: 25 days.
	• Long-time goal: 14 days.
Reinforce & celebrate	• Notes of appreciation were added to quarterly score cards, expressing thanks.
	• On quarterly sales calls to Purchasing, the sales rep drops by management's offices to say thanks.
	• Record-breaking sticker on chart for best-ever quarters.
	• Fast-food coupons for unloading team.
	• Fast food served at customer locations as goals met.
	• Model train hopper cars given as trophies.
	• For each car turned around in 14 days or less—one chance for a drawing for a weekend at the Chattanooga Choo Choo.
	• Dinner train excursion for management of successful customers at the end of the year.
Results	• From the initial conversations with customer management, they seemed unaware of the hold times and somewhat embarrassed. We made sure to not cast blame, saying we understood the complexity of the logistics, and that we were just asking to join forces to make improvements that would benefit both companies.
	• As soon as the data/feedback was shown, improvements began to happen and continued throughout the project as the process was improved, demonstrating that "Feedback is the Breakfast of Champions."

Railcar Hold Time

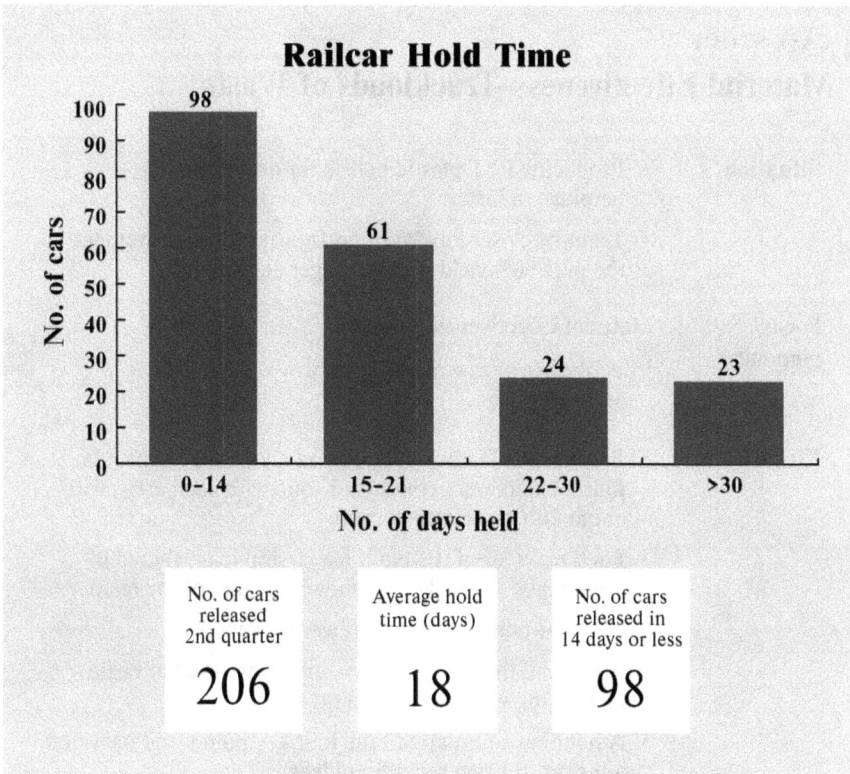

Figure 8.1. Railcar hold time feedback was given in quarterly reports.

CASE STUDY

Material Effectiveness–Truckloads of Waste

Situation	• Producing PET plastic pellets for use in the beverage industry.
	• Taking a closer look at opportunities for improvement, the over 90% yield was no longer acceptable.
Focus/ pinpoint	Material Effectiveness:
	Effective use of raw material improves yield and reduces waste.
Kickoff	• Because the factory operated on a 24/7/365 schedule, four kickoffs were conducted—one with each crew, with about 100 associates in each.
	• Each member of the plant leadership team shared in speaking at the kickoff to show the unity of the team.
	• The four points of a kickoff were covered.
	• As part of the kickoff, every team was asked to begin identifying what they could do to help.
	• A handout summarized the four key points and provided a means to begin recording ideas.
	• Some ideas for improvement were shared to prime the pump.
	• A microphone was provided, and employees were asked to share ideas.
	• Linking in was explained, and teams were asked to share their pinpoint/project with the leadership team at regularly scheduled link-in meetings.
Translate & link in	• At the weekly leadership team meeting, link-in teams shared their project ideas and were recognized and encouraged.
	• Over 40 teams linked in to the effort with formal projects.
	• The link-ins included not only teams from the production departments, but from all the supporting units as well, with every unit in the company eventually working on a project.

Management action	• The leadership team brought in Aubrey Daniels, management consultant, for a 2-day workshop centered around principles and concepts for a successful intervention.
	• Process-improvement resources were given to each team to help them in selecting their link-in and making the improvements.
	• The leadership team took on the challenge of finding a way to provide feedback on yield on a weekly basis and before the end of the 1st shift on Monday.
	• Weekly meetings were held to review performance and plan ways to interact with the workforce in sharing the results and recognizing progress.
Improve process	• Emphasis was placed on using best practices in all work areas.
	• Each work unit acting as an improvement team identified causes of upsets and then worked in teams to reduce or eliminate them.
	• This work took place in manufacturing, maintenance, materials handling, and support organizations.

Measure & feedback	• The measure was "percent right the first time."

• The measure was "percent right the first time."

• A scoreboard was designed for material effectiveness showing percent right each week, a monthly summary graph, best month ever, and year-to-date percent right.

• Percent right for the week was summarized, reviewed in the Monday leadership meeting, and posted on the plant-wide scoreboard.

• Comments on the scoreboard included positive reinforcement and notes for process-upset causes.

• A copy of the scoreboard was distributed to each work area to be displayed alongside their project scoreboard.

• At the end of each month, percentage right was added to the graph (Figure 8.2). The year-to-date average was updated, and if appropriate, the best month ever was updated.

• As the initiative progressed, at the suggestion of a factory operator, a bar graph was added to the scoreboard to show the equivalent number of truckloads of product not right the first time for each month.

• This addition to the graph became a rallying point, as the amount of waste was made more visible, tangible, and realistic. Operators and employees throughout the plant talked about the truckloads, how far they would line up down the road, and how it was embarrassing to work where that much of their production did not meet standards.

• There was talk about how much material had to be purchased to make a unit of good product. For example, to make 100 units of good product, enough raw material had to be purchased to make 111 units. A good term for that ratio would be "material effectiveness."

Reinforce & celebrate	• Comments and notes on the scoreboard from the company president served as good means for reinforcement.
	• To interact with employees and to increase attention to the material effectiveness pinpoint, the leadership team met employees in the parking lot as they arrived for work. They congratulated them on their progress and new records of performance. They asked employees how they were linked in, meaning what they were working on to bring about the needed improvement.
	• At one point, when they reached a new best-ever period, the vending machines in the plant were turned on free for a day.
Results	• From an initial baseline around 90%, percentage right the first time climbed to a level of over 99% over 7 years, steadily improving each year.
	• Truckloads of "not right yet" decreased from a starting point of 142 trucks per month to 1/4th of one truck by the end of those 7 years.

Resource Utilization

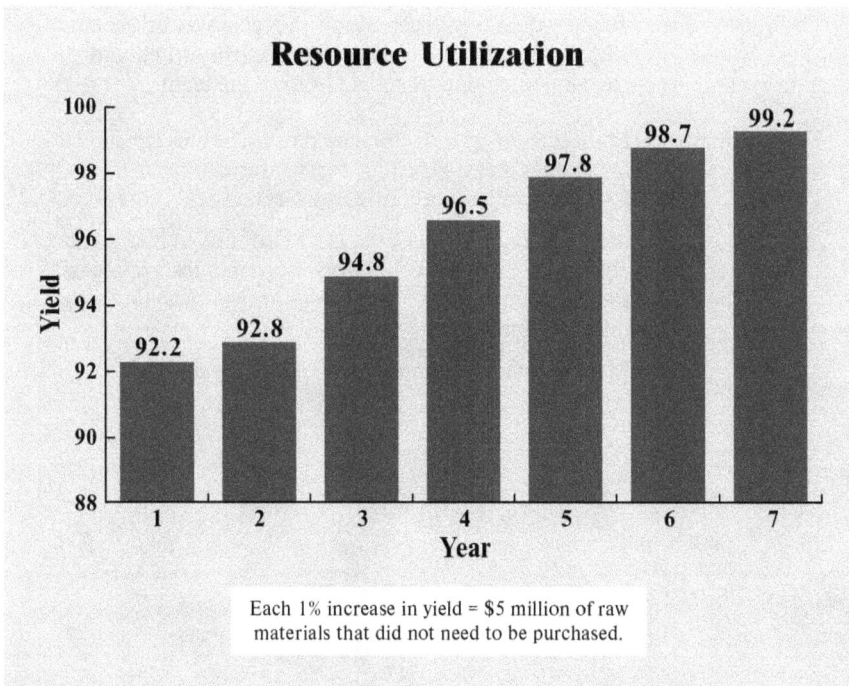

Each 1% increase in yield = $5 million of raw materials that did not need to be purchased.

Figure 8.2. *Resource utilization increased by 7% over 7 years.*

CASE STUDY

Make International Business Easy (MIBE)

Focus/ pinpoint	With the most potential for future growth in international markets and lower customer satisfaction scores there, the question was, "What are all the things we do that make business more difficult for our customers in other countries?" These were things such as custom documents, invoice errors, and exchange-rate problems. The challenge was to "make international business easy" (MIBE).
Kickoff	• 3-day workshops were held in each international office to explain the initiative and to begin developing improvement projects. • Kickoff meetings were also held with all departments impacting international business, such as Shipping, Production Planning, Finance, and Product Literature.
Translate & link in	• Over 100 teams linked in from more than 20 international locations, plus domestic supporting units. • Each link-in took place in a formal ceremony, where the team shared their pinpoint with the Marketing Presidents Team. At the link-in ceremony, the team was presented with a satellite to be placed on the MIBE scoreboard showing the pinpoint and a photo of the team.
Management action	• Management provided on-site process-improvement facilitators and training for every international office and for support teams initiating a project. • Management also designed the MIBE link-in board and conducted more than 100 link-in ceremonies. (Some of the ceremonies were held after hours to accommodate international working schedules.) • Management developed the MIBE Festival.

Improve process	A sampling of the more than 100 MIBE projects: • Delivery times were reduced by 45 days in Singapore by establishing local bulk storage. • Clean orders in Cologne increased from 56% to 98%. • Bank accounts were established outside the U.S. to make payments easier. • Response time to customer inquiries was reduced in Korea from 5 days to less than 3 days. • Printing of product brochures in local languages was increased. • Orders delayed for pricing were reduced from six per month to zero for 10 months in a row. • Order acknowledgement was reduced to less than 8 working hours. • All international document figures were converted to metric.
Measure & feedback	• As a team completed a project, a star was placed on their satellite on the MIBE scoreboard. • When the team initiated a second project (or more), they received a space shuttle for the scoreboard. • An international satellite at the top of the link-in board tracked the number of completed MIBE projects. Each project moved the satellite further along on its journey to circle the globe (40,000 kilometers) with improvements. • Each project team received a duplicate chart tracking team progress and a quarterly update of the overall progress. • A quarterly MIBE newsletter highlighted team achievements and new link-ins. • At the MIBE Festival, each team's storyboard was judged by a team of senior managers and given a quality score. The score provided feedback to the team on their success and provided learning for the judges. • Overall, the initiative was measured by customer satisfaction scores and sales revenue growth.

Reinforce & celebrate	• To celebrate the success of the MIBE initiative, an MIBE Festival was held at company headquarters. Teams from around the globe gathered for a day of celebration, to share their improvements, and just to get to know each other better. Over 900 people participated in the festival, which included international music, food, and attire. As project results were displayed on storyboards, participants traveled around the two-story employee center with their MIBE passport in hand to take notes on the countries and projects visited and ideas to take back to their own units.
	• As observers and exhibitors discussed the projects, the murmur eventually rose to a deafening, businesslike buzz.
	• At the lunchtime picnic on the grounds, languages from around the globe were spoken as teams shared information about their MIBE journey.
	• Festival attendees who suggested an idea for a new improvement project received three softballs to throw at the executive dunk tank.
	• A movie summarizing the MIBE effort and results, and recognizing the teams who had contributed, was shown in the employee center theater.
Results	• At the time of the first MIBE Festival (eventually there were three), 11 of the projects had been completed and the international satellite had traveled over 14,516 kilometers on its journey around the globe.
	• Within the second year of the MIBE initiative, over 130 teams had linked in and were on the board. Nearly half of the teams had completed their first project.
	• The second MIBE Festival was attended by over 1,200 people.
	• Along with the staggering results in the aim to make international business easy, with improved customer satisfaction and sales revenue growth, perhaps the most important impact of MIBE was to share the improvement ideas and lessons learned from around the globe and see them replicated in many other locations.

For more about this story see:

- MIBE – Make International Business Easy—PM Magazine, Vol 8, #4, page 19. © Aubrey Daniels International. Used with permission. https://www.aubreydaniels.com/media-center/ archive/performance-management-magazine-fall-winter-1990-vol-8-no-4

- Making International Business Easy (MIBE)—PM Magazine, Vol 9, #4, page 10. © Aubrey Daniels International. Used with permission. https://www.aubreydaniels.com/media-center/ archive/performance-management-magazine-fall-winter-1991-vol-9-no-4

CASE STUDY

On-Time Program Performance (OTPP)

Situation	• Design and manufacture data management and control systems for satellites, launch vehicles, and missiles. • These were exciting and very successful days for organizations in the space industry. • The best technology was winning the business, and customers gave a pass as far as on-time delivery was concerned, but those days were ending.
Focus/ pinpoint	On-Time Program Performance (OTPP) • Meeting scheduled delivery for systems • Baseline: 48%; Goal: 85% • Theme: OTPP will get us to Busch Gardens!
Kickoff	• In your workplace: Think what you can do to help meet schedules. • Form a team, identify a pinpoint, link in to the process, and share your project with the executive team. • Go solve the problem, and then celebrate.
Translate & link in	• Over 40 teams linked in. • The executive team was concerned about making the link-in teams voluntary but proceeded after being assured that only a few teams were needed to seed the effort. Then those first few would be reinforced, and interest would grow. 14 months later, 115 teams had linked in.

Management action	• Employees were asked, "What should be done?"
	• Every Friday, management had a meeting to listen to and reinforce the linked-in teams and to encourage them to make decisions about what they would do to improve OTPP.
	• PAThfinders (Performance Acceleration Team) was created to assist teams to identify potential improvements and to serve as advisors to the teams using their experience and continuous improvement training (in addition to their regular jobs).
	• Each PAThfinder was paired with an executive team member to focus on teams in that area.
	• The executive team wore safari shorts, hiking boots, and straw hats during link-ins and celebrations.
	• Calling themselves the Bureaucracy Busters, the executive team took a bite out of bureaucracy by eliminating time-consuming and unnecessary approval signatures—20,000 annually by the time the project was finished.
	• They reduced or eliminated a total of 150 reports and meetings.
Improve process	• Each work unit acting as an improvement team identified causes of delayed deliveries.
	• Focus on boosting on-time performance from vendors, Engineering, Production, Testing, and Administration.
	• Song: "Programs Need to Be on Time" (to the tune of "Happy Days Are Here Again").
	• With the delivery of two space-system computer units falling far behind on cost and delivery (and considering adding more people), a 911 team was formed and on-time delivery went from 11% to 87%, with cost also under budget.
	• Formed an 831 team for an important unit due August 31. For the first time in the operation's history, the team built the unit, inspected it, cleared the paperwork, and delivered the unit within 10 weeks, beating the schedule by 30%–40%.

Measure & feedback	• A 16' x 24' mural with a jungle motif was painted on the outside wall of the building. The mural was a map of three possible trails to get to Busch Gardens:
	
	   ○ 12 months with an average of 85% on-time delivery, with the longest route traveled by a turtle and 12 traffic lights to go through.
	
	   ○ 6 consecutive months of 85% or above, with six traffic lights and traveled by a rhino.
	
	   ○ 3 months of 100%, with three traffic lights traveled by a cheetah.
	
	• As key milestones were met, the company president used a bucket truck to reach the scoreboard and move the animals as they progressed.
	
	   ○ A large, plastic traffic light was displayed for each program: green if on schedule, yellow if in the caution zone, and red if in trouble.
	
	• For the bureaucracy buster executive-team project, a wall of bureaucracy was built from wooden blocks. Each time a signature was eliminated, a block came off the wall. At the end of the project, each team member received a brick with an engraved plaque saying, "Thanks for breaking down the walls of bureaucracy."
	
	• The company newsletter kept all employees informed on activities supporting OTPP and monthly progress.

Reinforce & celebrate	• In keeping with the safari to Busch Gardens theme, celebrations included straw pith helmets, zoo pretzels, Matchbox™ jeeps, bananas, and jungle punch as colorful trophy-value items for milestone events. • Busch Gardens was reached by the path of 6 consecutive months of 85% or above, and the day was celebrated by 4,600 people enjoying Busch Gardens Day for the company. The company vice president greeted employees and their families as they entered the gate. • The VP and leader of the OTPP initiative was honored by receiving the company's Chairman Achievement Award. • The CEO thanked the vice president for his extraordinary leadership. "You truly lead by example, and we are fortunate to have you on the team."

Results

- It took 8 months to reach 85% the first time, but once there, the workforce never faltered (see Figure 8.3). They reached the goal via the rhino path, with 6 consecutive months at 85% or more of on-time delivery. (Remember, they started at 48%!)

- All-time high during Phase 1: 96%.

- 115 teams linked in.

- From 21 programs in the green to 36.

- From 10 programs in the red to 4.

- Cost performance: 37% improvement.

- 20,000 signatures eliminated annually.

- 150 reports or meetings simplified or eliminated.

- Won the state Quality Team Showcase award.

- Next goal 90%—for a trip to "New Orleans" and a Mardi Gras celebration.

For more about this story see:

- The Sky's the Limit – Honeywell's Space and Strategic Systems Operation Blasts Off–PM Magazine, Vol 14, #4, page 16. © Aubrey Daniels International. Used with permission. https://www.aubreydaniels.com/media-center/archive/performance-management-magazine-fall-1996-vol-14-no-4

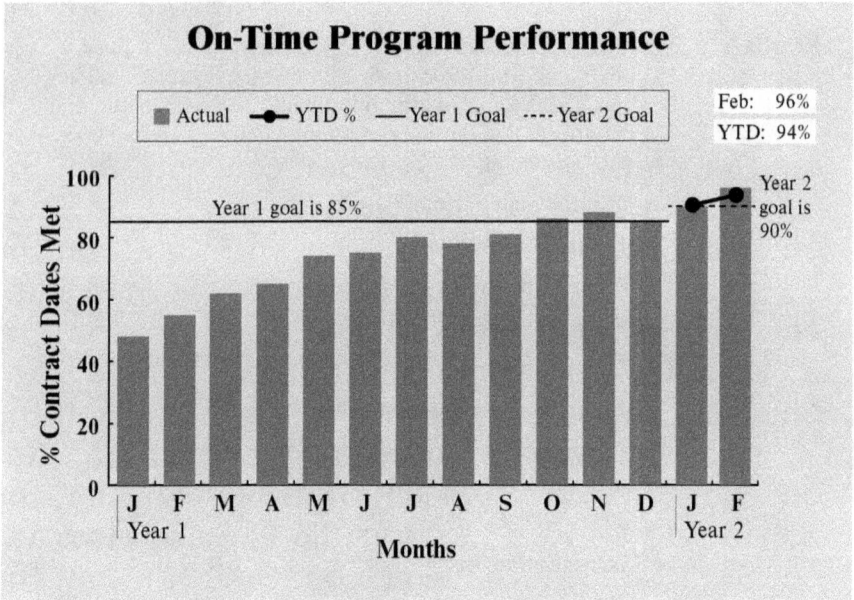

Figure 8.3. *Graphical display of the percentage of time contract due dates for deliverable hardware and software was met, indicating performance in satisfying customers' schedule requirements.*

CASE STUDY

Get Out and Stay Out (GOSO)–A Model for Rehabilitation in Corrections

Situation	• We are all familiar with the issue of recidivism–the tendency of a convicted offender to reoffend–and with the lack of success in reducing this tendency.
	• The most painful effect is the impact on the offender and their family.
	• In addition, there is an impact on society, as crime increases and the cost of incarceration rises.
	• Offenders across the state, while serving their sentence, have the opportunity to work in the woodworking facility, metal-working shop, data processing, dairy farm, and textiles. The statewide goal is to prepare offenders for success after release.
Focus/ pinpoint	• Increase the number of offenders who get out and stay out (GOSO).
	• Reduce recidivism.
	• Use a baseball theme of rounding the bases and heading for home.
Kickoff	Conduct Central Office rollout for staff at a baseball stadium.
	• Introduce the starting line-up.
	• Throw out the first pitch.
	• Have a famous player, coach, or announcer talk about teamwork, discipline, and the keys to winning.
	Why do we need to improve this?
	• Help offenders break the chain for future generations.
	• Help offenders realize their true potential.
	• Create a safer community.
	• Transform tax consumers into taxpayers.
	Who are the performers?
	• Offenders
	• All Department of Corrections employees
	• Other state agencies
	• Offender families
	• Legislators

Translate & link in	• At each offender work center, conduct a kickoff and create their GOSO Scoreboard.
	• Hold workshops at each work center to create an Accelerated Continuous Improvement initiative in the workplace.
Management action and improve process	• Develop an "offender profile" database to identify causes of return and ideas to prevent it.
	• Identify and enlist offenders to participate in this initiative.
	• Increase opportunities for diplomas and certificates.
	• Develop education & training workshops (life skills, coping skills, occupational skills, general education, computer training, jobs training).
	• Share success stories with staff and offenders.
	• Increase job opportunities by forming partnerships with felon-friendly organizations in offenders' home areas.
	• Increase Post-Release/Transitional Support System, including housing and transportation assistance.
	• Assign mentors (past offenders who have stayed out).
	• Share success stories. Bring back past offenders who have been out for 3 years or more (Hall of Famers), and have them speak in workshops.
	• Involve the family in pre- and post-release case management.
	• Increase community & faith-based support.

Measure & feedback	• Create a baseball scoreboard (Figure 8.4) in each prison work area (metal, wood, farm, clothing, etc.), which is a 4' x 8' sheet of plywood with a baseball diamond.
	
	• Establish criteria for the following:
	   ◦ Getting on the team (in the dugout).
	   ◦ Making the team and receiving a baseball card with photo on the front and stats (the accomplishments achieved in GOSO) on the back.
	   ◦ Advance to 1st base, 2nd base, 3rd base, and home.
	
	• Place "baseball cards" by the base (batter's box, 1st, 2nd, 3rd, home) representing the level of progress that has been made in getting out and staying out.
	
	• Hold weekly meetings "around the scoreboard" to move those offenders who have qualified to the next base and discuss how they did it.
	
	• Maintain a stay-out graph.
	
	• Each quarter, determine the number of offenders released 3 years ago and the number of those still out.
	   ◦ Post this success rate (batting average) on the stay-out graph.
	   ◦ Interview those who have stayed out and document their story.
	   ◦ Interview those who returned and determine the causes.
	
	• Update box score (Figure 8.4) monthly with (1) the number of offenders who were released 3 years ago and (2) the number still out. (Study each person for that month via interview to determine how they stayed out or why they came back.)
	
	• Publish success stories.
	   ◦ Give to current offenders, staff, and state officials.
	   ◦ In person, written, and video.

Reinforce & celebrate	Reinforcement for offenders and staff:
	• Give celebration snacks and meals as offenders run the bases.
	• Trading cards: collectible.
	• Class graduation celebrations.
	• Peanuts, popcorn, Cracker Jacks.
	• Success stories shared by past offenders.
	• "Take Me Out to the Ballgame" song with each person still out for 2 years (7th-inning stretch).
	• On the anniversary date of a person being out for 3 years, ring a bell three times and hear their story.
	• Hall of Fame for offenders who GOSO 3-plus years.
	Reinforcement for staff:
	• Tickets to a baseball game.
	• Autographed baseball.
	• Baseball caps.

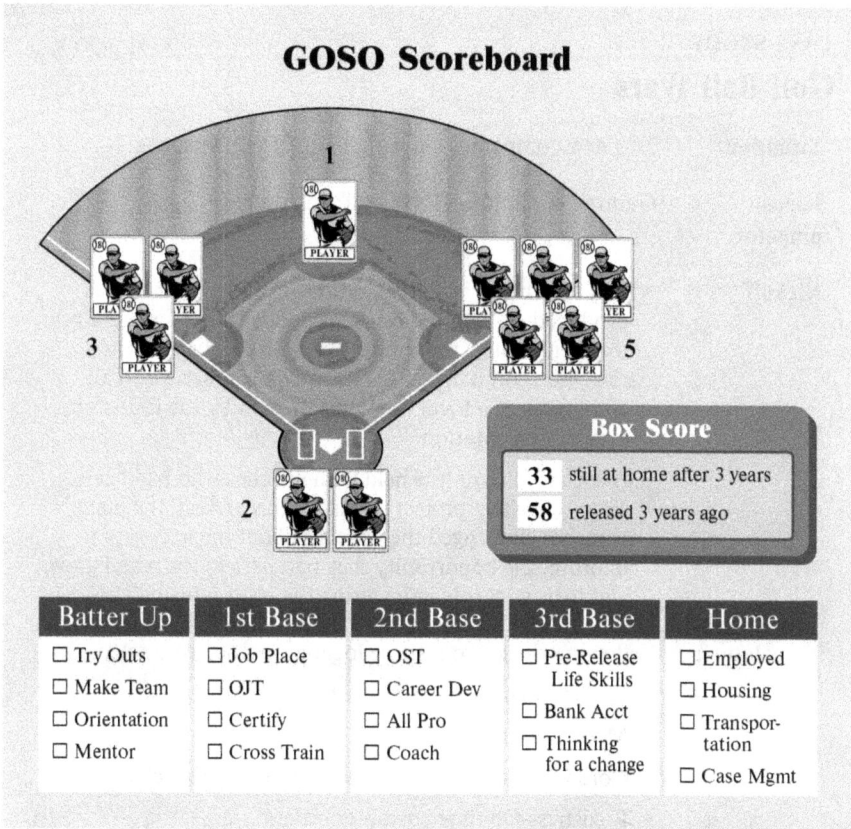

GOSO Scoreboard

Box Score

| 33 | still at home after 3 years |
| 58 | released 3 years ago |

Batter Up	1st Base	2nd Base	3rd Base	Home
☐ Try Outs	☐ Job Place	☐ OST	☐ Pre-Release Life Skills	☐ Employed
☐ Make Team	☐ OJT	☐ Career Dev	☐ Bank Acct	☐ Housing
☐ Orientation	☐ Certify	☐ All Pro	☐ Thinking for a change	☐ Transportation
☐ Mentor	☐ Cross Train	☐ Coach		☐ Case Mgmt

Figure 8.4. Progress "running the bases" is depicted on the GOSO scoreboard.

📁 CASE STUDY

Golf Ball Wars

Situation	Two new companies entering the golf ball business
Focus/ pinpoint	Gaining & retaining (golf ball) customers
Kickoff	• To kick off the initiative, a *Golf Ball Wars* movie was produced. It featured a lake scene with the leadership team emerging from under water in full battle gear— ready for golf ball wars. The movie was shown to all employees at a local theater, followed by the four-step kickoff presentation. • A second kickoff was held with coaches who had been selected to help project teams. In this kickoff, the plant manager challenged the coaches about the necessity to improve, the opportunity they had here to learn and grow, and their vital role in inspiring the project teams.
Translate & link in	Following the kickoff, these local pinpoints were selected: • Packing—Bulk-pack miscounts • Molding—Changeover time • Core Components—Lost production due to lack of cores • Logistics—On-time customer orders • Stamping—Color consistency • Painting—Dirt-related defects
Management action	• A "Crackerjack Team" (CJT) composed of members of the leadership team, supervisors, and managers was appointed to oversee the development and implementation of the ACI plan. • In addition to working sessions, CJT meetings were held monthly to conduct link-in ceremonies. At the link-ins, the teams presented their plan to the CJT, explaining how their project linked to the overall company pinpoint, the value of reaching the goal, and some of the actions planned. This was a time for celebration and for encouragement by the CJT. • Coaches were established to help linked-in teams, provide resources as needed, and encourage the teams.

Improve process	• At the kickoff, the idea of Job #2 (finding a better way to do things tomorrow) was explained.
	• Crews were asked to work on improvement by finding root causes, developing ideas to prevent them, tightening up good practices, generating innovations, and keeping score.
Measure & feedback	A scorecard was developed to track progress.
	• Like a golf card, the scorecard had nine holes to play, each with a scale for determining the score on that hole.
	• The nine holes represented the critical factors for gaining and retaining customers.
Reinforce & celebrate	• Of course, for reinforcing and celebrating progress, milestones, and reaching goals, there are plenty of golf images to create interest: food and drink like ParBQ (barbecue), greens (turnip), and tee (sweet tea); and golf shirts, tees, logo balls, and golf trading cards.
	• As the consultant, I was thanked with custom logo golf balls for my son's wedding and for my men's group at church.

CASE STUDY

Triple Crown

Situation	• 1,700 stores are served by 14 Regional Distribution Centers (RDCs) throughout the U.S., with each RDC serving approximately 100 stores with a dedicated dock for each store.
	• The challenge is to get the right, undamaged replacement items moving toward the stores as soon as possible after a purchase takes place.
	• This is done by loading only the correct items on the truck and packing it "high and tight" to reduce damage and freight cost.
Focus/ pinpoint	Moving beyond the Boss Award (only one winner) to the Triple Crown (where every RDC could win)
	Three focus areas all point toward exceeding customer/ store expectations:
	1. Shipping accuracy (getting what the store wants)
	2. Shipping quality (getting it in good condition)
	3. Shipping cycle time (how long it takes to get what the store wants)
	Theme ideas: "stop shipping air," "it can't be sold while it is still in the warehouse," and "high and tight."

| **Kickoff** | Conduct a kickoff in the warehouse assembly area, with all associates present: pickers, loaders, scanners, runners, planners, and coaches. Specific activities: |

- Hang a screen for a slideshow.
- Have Spanish translation on slides.
- Have refreshments.
- Have "customers" (store representatives) share information about the importance of cube, audit, and cycle time.
- Show a "news story" video of what it's like at a store when unloading a truck—one excellent truck and one with problems.
- Have a card for each person to fill out with ideas on how to solve problems and good practices related to cube, audit, and cycle time.
- Employees turn in cards with ideas on the way out, to be entered in a drawing for free movie tickets.
- Cross-functional team summarizes cards and shares with all.

Discuss why we are launching this initiative:

- Stores and customers get what they want, when they want it.
- An increase of 100 in cube (the amount of space filled) reduces shipping by $300,000 per year.
- Damage is reduced.
- Stores are more productive.
- Replacement shipments are reduced.
- Less inventory in stores with faster cycle time.
- Items mis-shipped have to be handled extra times and examined to determine what they are and what to do with them. New orders have to be cut, and records have to be updated.

Goals: Cut cycle time from 122 hours to 97, audit from 10 errors to 5, and increase cube from 2,850 to 3,100.

| **Translate & link in** | See Improve Process |

Management action	• Form three teams, one for each focus area, to finalize plans and manage progress.
	• Create a Scorekeeper position for each shift on the floor to make sure the score is posted when an event occurs—truck loaded and closed, incoming truck unloaded, finish a picking packet. The Scorekeeper is a member of the operations team.
	• Summarize the list of audit errors and post it on the "Help Wanted" board.
	• Hold one inspection per week for each bay/dock to evaluate the quality of the cube and to identify good practices. Inspection conducted by a coach/lead/manager **and** pickers/throwers/closers/loaders. (They learn more from inspecting than from being inspected.) Good practices identified are discussed in shift/team meetings.
	• Create a Damage Committee to make sure every damaged item is investigated for root cause and to appoint teams to work on prevention.
	• Hold "High and Tight" training and practice sessions on increasing the cube.
	• Create a Permanent Fix Club for individuals or teams developing a fix to an ongoing problem.
	• Make English as a second language classes available to all employees at no charge.
	• Schedule every coach, lead, loader, and planner to go on an audit.
	• Hold an annual Loading Competition.

Improve process	• Shipping accuracy (audit)
	    ◦ As the list of audit errors—(1) item on invoice, not on truck (under); (2) item on truck, not on invoice (over); and (3) damaged items—is posted on the Help Wanted board, individuals and teams work to find root causes and solutions to the causes of errors.
	
	• Shipping quality (cube)
	    ◦ Shippers/runners analyze inspection reports and generate ideas for improving the cube.
	
	• Shipping cycle time
	    ◦ Tabulate the frequency and causes of "potholes" (the little things that slow us down) like mislabeled items, damaged items, misplaced items, and items scanned but not loaded. Beginning with the most frequent, appoint teams to tackle the causes.
	
	• Create an Equipment Uptime Team to focus on conveyors, stackers, and computers.
	
	• Create a vendor scorecard—to measure labels pointing out, similar product together, correct count, and damage.

Measure & feedback	• Cycle time: baseline 122 hours, goal 96 hours.
	• Cube: baseline 2,810, goal 3,000.
	• Audit: baseline 10, goal 5.
	• Determine shipping accuracy and post the score and a list of the shipping errors on the dock-feedback sign. Summarize the list of causes and post on the Help Wanted board.
	• Determine shipping quality:
	    ◦ When a load is completed (and door to be closed), the lead/coach posts cube score on the dock-feedback sign with a photo of the load.
	    ◦ Calculate the number of truckloads eliminated by increased cube score.
	• Put up a scoreboard (see Figure 8.5) in the assembly area:
	    ◦ Post cube score each day.
	    ◦ Audit score when received.
	    ◦ Cycle time each week.

Reinforce & celebrate	• Fly the "perfect shipment" banner for 100% truck audit.
	• Cube scores over 3,000 turn on the good-cube light.
	• Weekly average of 3,000 cube score: Serve refreshments with different shapes and colored ice cubes.
	• Monthly average of 3,000 cube score: Distribute photo cubes with pictures of high and tight truckloads.
	• Rubik's Cube® given to everyone with a special RDC photo on each side.
	• Celebrate a best ever in any category with a free lunch.
	• Triple Crown for a quarter: Throw a tailgate breakfast in the parking lot as employees arrive.
	• Triple Crown for the year: Have a family picnic with "cube" steaks grilled by supervisors and management.
	• Hold a vendor banquet for those with top scores.

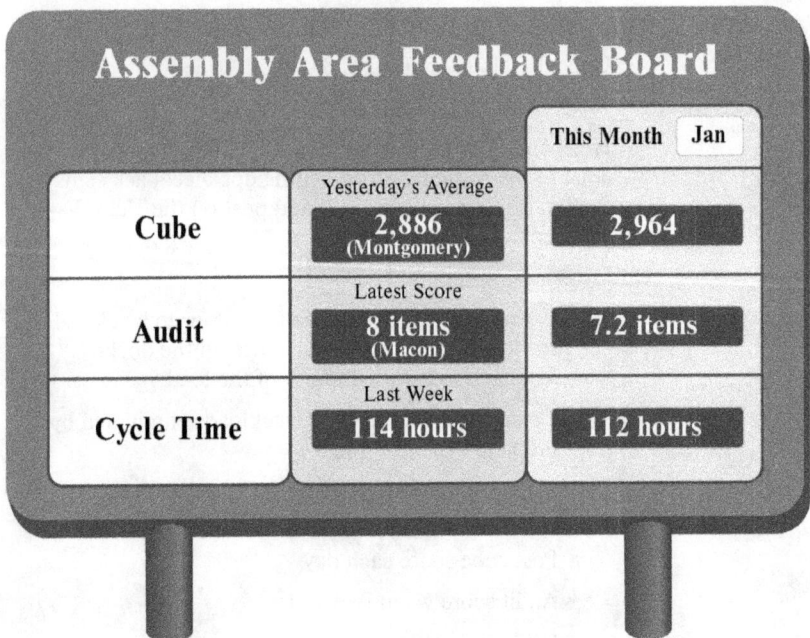

Assembly Area Feedback Board

This Month Jan

	Yesterday's Average	
Cube	2,886 (Montgomery)	2,964
	Latest Score	
Audit	8 items (Macon)	7.2 items
	Last Week	
Cycle Time	114 hours	112 hours

Figure 8.5. The Triple Crown scoreboard keeps everyone informed of the progress in better serving the stores.

Wrapping Up Success in the Gift Wrap Factory

One of my most successful and enjoyable consulting and coaching experiences took place in Greeneville, Tennessee, at the gift wrap factory there. A group of employees from the factory attended an ACI workshop that I had conducted at the Pal's Sudden Service training center. Some of those employees went back to the plant and began applying the principles from the workshop to their respective operations, with noticeable results. When Scott Crawford, the new plant manager, arrived, some of the employees convinced him to attend an ACI workshop. He arrived at the workshop an admitted skeptic, but I could tell by the end of the 3 days that he was getting it.

I followed up with Scott after the workshop, and he was excited to put this new approach to leadership and improvement into place. Like few other leaders I have worked with, Scott demonstrated the ability to inspire others to put their shoulder to the task and follow his lead, while at the same time, giving them much freedom to determine the details and be innovative.

Working with a group that Scott assembled, the ACI plan was developed. After a good bit of wrestling, the focus of "feet of wrapping paper produced per factory dollar spent" was agreed on (achieving like-mindedness). The scoreboard was designed, the reinforcement planned, and a "racing" theme selected to give a face to the initiative. Work units throughout the factory and in administrative areas linked into the focus of "feet per dollar."

The results were phenomenal (17% improvement in feet per dollar), exceeding the expectations. There was some sentiment that "we need to take this ACI approach to headquarters; they really need it." But at my insistence, we said "no, we want to produce such good results that headquarters will be coming to us to see what is going on." And that's what eventually happened, as interest spread around the company at other factory locations and at corporate. Scott organized a workshop in Nashville, Tennessee, for plant managers and key support staff. As a result, the ACI approach began to spread contagiously around the company.

For me, it was very rewarding and encouraging to see the effort that was put into accelerating improvement and the results throughout the company. The opportunity to work alongside a host of leaders and support staff in the company was a real blessing.

See McClaskey Excellence Institute (2010) for the rest of the story.

Now for *You*!

- If you keep doing what you are doing, you will keep getting what you are getting.
- Your system is working perfectly today, just the way it is designed.
- You must change your world (see Figure 8.6).

Your return on investment from this book is
entirely dependent on what you **do** *now.*

Changing *Your* World

1. Reinvent the annual/strategic planning process. Pinpoint/focus on *the* critical issue(s) for future success.
2. Align and mobilize the workforce around the pinpoint/focus (linking-in).
3. Establish the notion of two jobs for every associate as the working norm: Best-known way today; better way tomorrow through prevention and innovation.
4. Make performance visible. Use scoreboards and scorecards to turn measures into feedback.
5. Create a culture of celebrating milestones and results and reinforcing the behaviors that led to the milestones and results.
6. Become an on-the-floor and in-the-office coach.
7. Learn by doing. Learning is over. Time to *do*.
8. Stop some things being done now to begin leading this way.

9. Vote with your calendar. Your calendar is the real indicator of your priorities.

10. If not this approach, *what?*

New worldview. New leadership approach. New job. → *Accelerated Continuous Improvement*

Your Scorecard

Accelerating Continuous Improvement Scorecard—
For the 21st-Century Leader

Responsibility/indicators	Score F	P	H
Pinpoint (Focus the organization—simplify & concentrate) • Customer focused • Simple (any associate can explain) • Rally-able (associates would brag about improvement) • Unifying theme			
Linkage (Mobilize and align the workforce) • Compelling case for change can be explained by all employees. • Work teams have clear linked project (on link-in board). • Time is set aside for regular process-improvement team meetings. • Associates can answer the four questions.			
Measurement/feedback (Make performance visible) • Easy to measure—any associate can do it. (Count is best.) • Baseline, goals, and best evers are shown on a graph. • Associates know their team score at the end of the day. • Organization scoreboard passes the walk-by test. • All graphs are up-to-date. (If not, take them down.)			
Goals (Provide opportunities for reinforcement) • Against self and past—not against a standard or other company units. • Emphasize shaping—recognizing small improvements. • Minimize use of deadlines.			
Process improvement • Employees say they have two jobs: best-known way today, better way tomorrow. • Problems are identified, investigated, and eliminated. • Innovative improvements are implemented and sustained.			
Reinforce behaviors & celebrate results • Milestones, best evers, and strings of success are celebrated following the four steps. • Behaviors leading to results are recognized and reinforced.			
Coaching: Out of the office and onto the floor • Practice out of the office and into the gemba on a regular basis.			

Figure 8.6. *Note. F = fail; P = pass; H = honors. For passing and honor levels, data must be provided (e.g., number of teams linked in, number of projects completed, or score on celebration).*

Results to Expect

- 15%-50% performance improvements
- Commitment instead of compliance
- Make it work instead of did my part
- Best ever instead of standard
- Teamwork instead of individual
- Cooperation instead of competition
- Celebrations instead of reprimands
- Way of life instead of program

Appendices

More Pinpoint Examples

X-Ray Repeats—A 98% success rate on x-rays seems good ... unless you are one of the 2% of patients who must have your x-ray re-taken, and possibly have to go back through an uncomfortable preparation process.

Perfect Patient Visit—How many patient visits are perfect? To be perfect required successful performance on 18 points, beginning with answering the phone within three rings and ending with receiving payment within 30 days.

Engineering Yield—What percentage of entering freshmen in engineering graduated in their chosen field? A "yield" of 50% would put you out of most any business. Why should it be any different here? We were "wasting" students because of lack of readiness, roommate problems, health issues, and so on. Yield had to be improved.

Wait Time—Emergency room wait times—enough said!

Cows in the Sick Lot—One of the key factors for maintaining milk production on a dairy farm is keeping the cows well and out of the sick lot.

Frequent Diners—Flash cards were memorized by every employee, for each frequent diner, showing their photo and name, and their favorite foods, drinks, and table. By the way, frequent diners tip more.

Patient Delight—Go beyond satisfying patients to delighting them through minimum wait time, fast recovery, no reoccurrence, and easy handling of insurance.

Sudden Service—In the quick service restaurant industry, better known as the fast-food industry, what could possibly be more important than "time to serve"?

Ride the Wave—In a greeting card sorting operation, orders are combined into what is called "a wave." When the picks per hours in the sorting operation are humming, we are "riding the wave."

Ready to Roll—Have all assigned trucks/trailers ready to roll out on Monday morning for their assigned task to repair concrete, coatings, manholes, and leaks.

Time to Efficiency (TTE) Installation of New State-of-the Art Equipment—Changing the entire operation by installing the latest, most sophisticated equipment in the industry was proving to be more of a challenge than expected. Installation ran into lots of snags and took much longer than planned. Going forward, a challenge and a new theme were needed. Solution: Focus on the number of days until the new equipment is running smoothly according to specifications. Emphasis was placed on anticipating problems and delays and heading them off, while capturing and sharing good practices that moved the project along faster.

Health, Not Growth—Pinpoints often tend toward growth of sales, profits, and territory. Another slant is to focus on health. All healthy things grow. By focusing here, you move back in the chain from results to the process.

World-Class Managers—With over 700 fast-food locations across Canada, there is a need to standardize the job of store manager and to replicate good practices. To that end, the establishment of a World-Class Manager Certification becomes a tool. To increase opportunities for recognition, certification levels should exist. A map of Canada showing where World-Class Managers are located tracks overall progress.

Time to Report Test Results—In a hospital or doctor's office situation, mental health can often be just as important as physical health. Actually, they are related. How long it takes to get test results contributes greatly to anxiety, frustration, and stress.

Loan-Cycle Approval Time—Banks work to decrease the amount of time between loan requests and approval.

Clean Starts—In a pharmaceutical manufacturing operation, does the first batch produced meet quality specifications?

Batting Average–This represents the number of chemical product samples sent to customers that lead to orders and future business.

THE STORY

The Fat Rabbit

While I was leading a workshop in Japan and talking about focus, one of the participants in the workshop spoke up and said, "You people in America don't know how to catch rabbits." Puzzled by his comment, I asked him to explain further. He elaborated with this: "You go out rabbit hunting; spot a rabbit; get a bead on him with your gun; and just about the time you are ready to pull the trigger, you see another rabbit hop up that looks bigger to you. So, you change your aim toward the new, fat rabbit. Just about the time you are ready to shoot the fat rabbit, you see a white rabbit pop up and think, 'Oh my, I would rather have a white rabbit than the gray one.' So, off you go hunting down the white rabbit. All you Americans do is chase rabbits; you never catch one." He went on to say, "In Japan we don't chase rabbits, we catch rabbits. Once we choose a rabbit, we pursue it until it is caught. Once it is caught, you have earned the right to look around for another rabbit." He said further, "Justice-san, I know what you are thinking. You are thinking, 'Yes, but you didn't get the fattest rabbit.' Maybe so, but at least we have a rabbit instead of just chasing rabbits all the time."

More Creative Feedback Examples

Climbing the Sales Revenue Mountain

With a workforce of thousands of people in more than 20 countries, how could we represent sales in a meaningful and easy-to-understand way?

Since company sales were more than $1 billion each year—an amount difficult to comprehend—we decided our sales graph would be shown in million dollars per day. Shown on the graph was the history of the past 2 years—$6.5 million per day, $7.1 million per day—and the goal of $7.5 million per day for this year. We explained the magnitude of these numbers as compared to the cost of a new hospital, school, or church—an amount we could all grasp.

Just past mid-year, as the line graph being used began to run off the top of the scale (Figure B.1), the leadership team wanted to change the scale on the graph and put out new charts. I absolutely refused, saying that the workforce had run us off the top of the graph and they deserved to see it that way.

How many graphs do you have where, even with a challenging goal, performance shoots right past the goal and off the graph? Don't accept anything less. In this case, it happened by rallying thousands of people around the globe to find their part in driving sales revenue.

As the end of the third quarter approached and it was clear we were going to far exceed our goal of $7.5 million/day, we wanted a way to emphasize this and let our people revel in the accomplishment. The new challenge became to climb to higher and higher levels year after year. To represent this climb, we used mountain peaks (see Figure B.2). Last year's peak and the best-years-ever peaks were shown. A team of mountain climbers representing the people of the company was shown climbing the peaks.

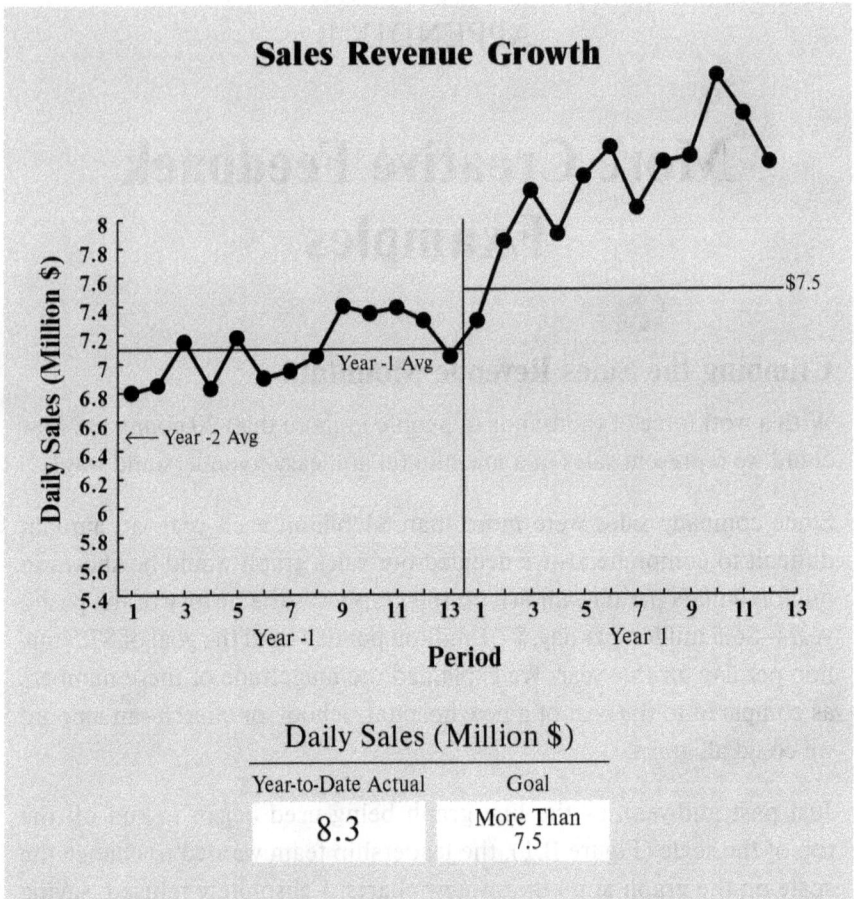

Figure B.1. Sales revenue ran off the chart.

Marketing Team Visualizes Sales Revenue Growth

What better way to measure sales than by updating the scoreboard when the order is entered into the system (when the customer submits the order, or the order correspondent pushes Enter)? And what better location for the scoreboard than in the Marketing Building cafeteria? And what better type of scoreboard than a flipboard (like you may have seen in international train stations)—with the flipping noise and action? The top 10 orders for this month, in descending order, can be shown on the board.

Imagine this! The Marketing cafeteria is full at lunch time with sales reps, order correspondents, and management personnel. A call comes in for an order that is the 2nd biggest this month. Suddenly, the board starts

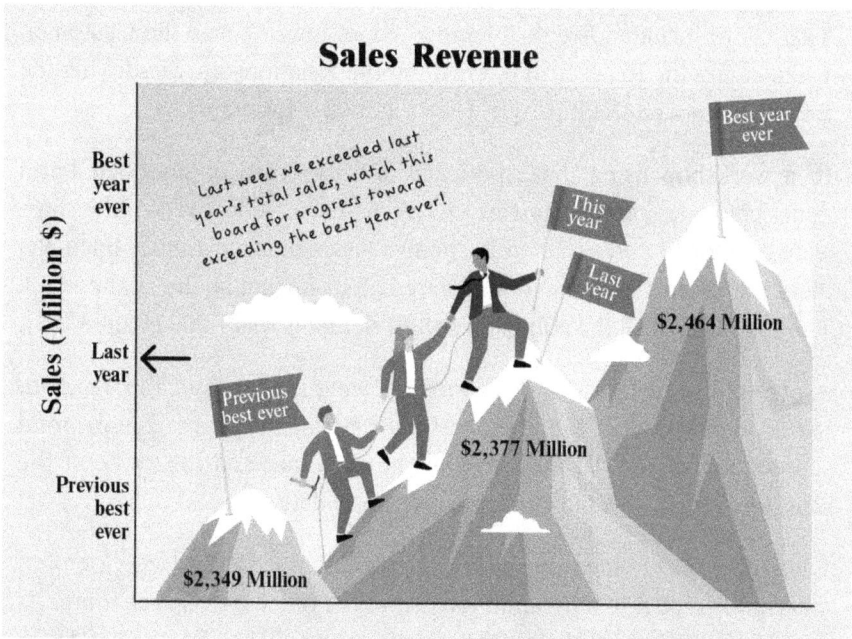

Sales Revenue

Sales (Million $)

Best year ever

Last year

Previous best ever

Last week we exceeded last year's total sales, watch this board for progress toward exceeding the best year ever!

Best year ever

This year

Last year

$2,464 Million

Previous best ever

$2,377 Million

$2,349 Million

Figure B.2. Reaching each higher mountain peak (sales revenue level) was recognized and celebrated.

flipping, and a new order with the amount and the customer moves into second place on the board. At one table, someone ask one of the sales reps, customer service reps, or order correspondents sitting there, "Isn't that your customer that just moved into spot number two?" Congratulations follow around the table, and the question is asked, "How did you do it?" The telling of the story begins.

Others walk in at lunch or during the day and see the board. Similar scoreboards on computers are available for remote locations.

Table-Cover Waste

When making paper table covers for picnics and parties, waste is generated by jams, paper humidity, and so on. Measuring this week's waste as how much of a football field would be covered is both meaningful and rally-able.

Climbing the Pyramid of Customer Relations

The Latin American sales team decided that the relationship-building process could be well represented as climbing a pyramid with the customer.

That climb included five distinct levels: Getting to Know the Customer, Establishing the Relationship, Growing the Relationship, Solidifying the Relationship, and the Pinnacle. (See Figure 2.1 for graphic.)

In a workshop using the experience and expertise of seasoned Latin American sales representatives, 30 elements for effective relationships were developed and assigned to the five levels of the pyramid. Each element was given a numerical value based on its impact on the relationship. The sum total of all the points for the 30 elements was 1,000 (Figure B.3).

A pyramid scorecard was developed showing the levels of the pyramid, the 30 elements, and the points assigned to each element. The pyramid scorecard also showed the name of the customer and the name of the sales rep, along with the cumulative score to date.

Each office was to identify key customers and place a flag (with logo) on a five-layer model of a pyramid located in the office conference room. At weekly meetings, completion of action items would be discussed and tabs for completed elements torn off. As customers accumulated points, the customer-logo flag moved up the pyramid to the proper level illustrating the relationship with the customer. Successful moves to a new level were discussed and celebrated.

Cost to Deliver

Like the cost of a postage stamp to deliver a letter, what is the cost to print, cut, box, and deliver a greeting card to the customer?

Print a large stamp each week with that week's cost displayed.

Railcar Spotting

Receiving over 100 railcars per day presents a challenge to get all of them in the right spot. Hopper cars of coal for the Power Plant are not so much a problem, but tank cars must be spotted at specific locations— locations where the piping is designed for unloading that chemical without contamination.

To solve this problem, Jerry, the engineer working on this project, developed a simple visual-representation card that showed the railcar that was

Strengthening Customer Relationship

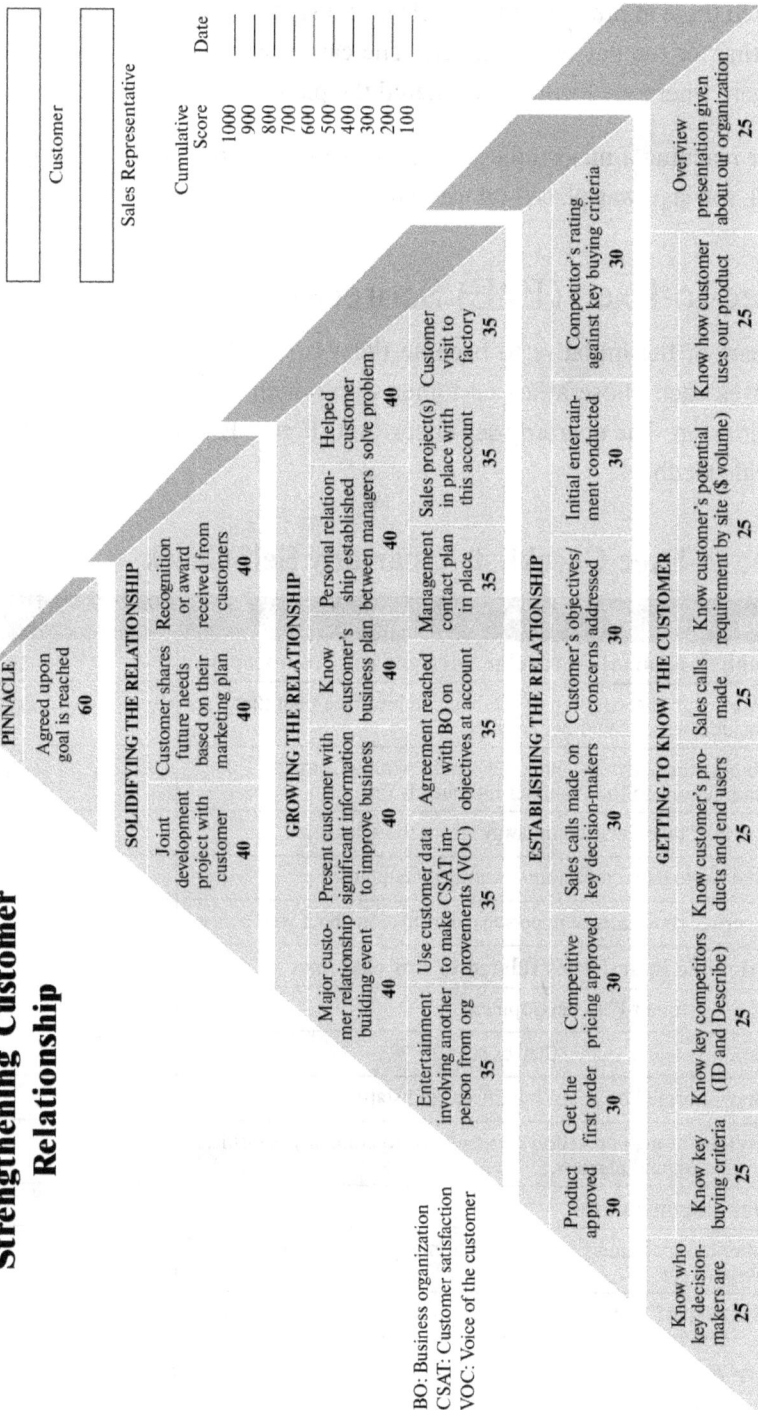

Customer _____

Sales Representative _____

Cumulative Score | Date
1000
900
800
700
600
500
400
300
200
100

BO: Business organization
CSAT: Customer satisfaction
VOC: Voice of the customer

PINNACLE

Agreed upon goal is reached
60

SOLIDIFYING THE RELATIONSHIP

Joint development project with customer	Customer shares future needs based on their marketing plan	Recognition or award received from customers
40	40	40

GROWING THE RELATIONSHIP

Major customer relationship building event	Present customer with significant information to improve business	Know customer's business plan	Personal relationship established between managers	Helped customer solve problem
40	40	40	40	40

Entertainment involving another person from org	Use customer data to make CSAT improvements (VOC)	Agreement reached with BO on objectives at account	Management contact plan in place	Sales project(s) in place with this account	Customer visit to factory
35	35	35	35	35	35

ESTABLISHING THE RELATIONSHIP

Product approved	Get the first order	Competitive pricing approved	Sales calls made on key decision-makers	Customer's objectives/ concerns addressed	Initial entertainment conducted	Competitor's rating against key buying criteria
30	30	30	30	30	30	30

GETTING TO KNOW THE CUSTOMER

Know who key decision-makers are	Know key buying criteria	Know key competitors (ID and Describe)	Know customer's products and end users	Sales calls made	Know customer's potential requirement by site ($ volume)	Know how customer uses our product	Overview presentation given about our organization
25	25	25	25	25	25	25	25

Figure B.3. The scorecard showed all the elements for building customer relationships and growing the business.

located there at present and the railcar that should be located there when spotting for the day was complete. The card was placed in a mailbox at the gate where the locomotive entered the plant.

After reaching a major milestone of days in a row of perfect spotting, a thank-you sign was placed on the gate.

Exec-to-Exec (E2E) Scorecard

As part of the initiative to become the World's Preferred Supplier, executives each chose a key customer with which to develop a personal relationship. The quarterly self-check tool (Figure B.4) was developed to provide feedback.

Exec-to-Exec Checklist—Quarterly Self-Check

Area	Fail	Pass	Honors
Establish & maintain regular interface with counterpart.			
• Work with the sales rep to establish an interface plan that includes visits and other contact.			
• Report the activities, outcomes, and learning from all interactions to the customer interface team.			
• Do not "take the account away" from the sales rep.			
• Stay up-to-date on customer's strategy & plans.			
• Learn & show interest in personal information about your contact.			
Participate in activities related to this customer			
• Key Customer Plan development.			
• Solution teams (when they exist).			
• Stay informed on goals, initiatives, and status.			
• Follow through on action items related to company priorities, resources, and barriers.			
Other responsibilities			
• Maintain continuity with this customer. This is a career-long assignment.			
• Back out of relationships where another executive is assigned.			

Figure B.4. The executives were pleased to have a clear definition of expectations.

Notes:

1. Participants in the Exec-to-Exec process were volunteers.

2. Participants were expected to complete a 2- to 4-hour workshop for training. This workshop included a review of responsibilities, exercises to practice, tips/watch outs, and demonstrations.

More Social Reinforcement Examples

Book of Records

Like the Guinness World Records™, maintain a book with a page for each key result measure, with entries showing the history of new records and the dates each time a new best ever is established. A photo of the team responsible for the new record can be included. A celebration event is called for each time a new entry is made in the book.

Beat the Budget Bandits (Basketball Theme)

When you reach a milestone, have a dribbling contest, organize a free-throw contest, play a game of HORSE, or show a Coach John Wooden film or a basketball all-time highlights reel.

The Ham Is in the Pan

Haynie, the department head, promised the crew of the Jet Building Shop (jets are perforated metal plates) that when they achieved 4 weeks in a row below the jet-failure reduction goal, he would provide ham biscuits for everyone. When 2 weeks in a row was met, he announced that he was purchasing the ham. After 3 weeks below the goal, he announced that "the ham is in the pan." When the 4-week goal was met, he showed up in a chef's hat with ham biscuits for all. One of the operators who was on vacation that day commented that he would always regret not being at work that day to see Haynie in that hat.

Postgame Tailgate Party

After a textile factory increases the yards of good fabric produced to 20% above last year, a postgame tailgate party is conducted in the parking lot, with a legendary sports announcer running the show. The "big plays" are recounted while everyone enjoys a tailgate cookout.

Training Ride

This reinforcement was designed for employees who had redesigned the "TRAINing process" and taken the madness out of it. They were treated to a scenic train ride day trip: Signs at the boarding location said "Chaos," and at the destination "Harmony," indicating the journey the team had taken.

M-Power-Mint

After the sales-support staff was successfully empowered to handle calls themselves rather than pass them up for a decision, they held an empowerment celebration with plenty of M&M's®, PowerAde drinks, and York peppermint patties.

Ready to Roll

Symbols for celebrating progress on having all maintenance trucks ready to roll out for the week's assignment on Monday mornings include Rolo® candy bars, a pair of dice, toilet paper rolls, cinnamon rolls, and roller coaster rides at a local theme park.

You Nailed It

The Nail Award was given to individuals or teams along with verbal thanks for solving a problem. These aluminum foil nails were sometimes as long as 6 feet.

More Tangible Reinforcement Examples

Football Stickers

We've all seen the lengths to which a football player will go—the extra effort invested—to get a sticker on their helmet. Similar stickers are equally effective for hard hats, work caps, and uniforms. They are the symbols that represent success and excellence.

Set the Table for a Feast

As a way to measure and reinforce progress when seeking to reduce table-cover (used for picnics and parties) waste, place a nice table in a visible location on the factory floor and "set the table" with one item for each week the goal is met. Progressively, add plates, utensils, glasses, napkins, creamer, sugar, and flowers. When the table is fully set, everyone enjoys a lunchtime feast or a family-night banquet.

Popsicles

Popsicles are given for preventing a "meltdown" of any sort, such as machine breakdowns or late orders.

Superhero Comic Books

When Shops and Services reached on-time service goals, superhero comic books (including some collector editions) were distributed to all team members. They enjoyed trading comic books and looking for the collector editions.

Push-Ups

Push-Ups® frozen dessert pops were given for increasing battery uptime for tow motors.

Football Trading Cards

When university students started receiving one-of-a-kind football trading cards for getting grades of 80% or more on tests, the grades skyrocketed from a 72% average to an 84% average. And the students actually started asking for more tests.

Reese's Peanut Butter Cups

Partnered with company leadership, the **PAT**hfinders (Performance Acceleration Team) assisted work units in identifying a focus and completing projects.

To recognize the work of the **PAT**hfinders and show appreciation, senior leaders personally delivered Reese's peanut butter cups to each **PAT**hfinder. A simple sticker on the back said, "Thank you to the **PAT**hfinders! The greatest partnership combination since chocolate and peanut butter!!"

Bonanza Bucks—Justice Family

We bought toys from the annual Bargain Bonanza thrift sale conducted by the Junior League. Then we printed up our own currency—bonanza bucks.

We put the toys in paper sacks and "priced" the toys based on their value to our children. (For example, 50 bonanza bucks for an American Girl doll for daughter Lauren.)

Son Matthew (and his friends) could earn bucks by doing chores listed on the refrigerator door. This not only stopped his complaining about chores, but he even asked me to put more items on the list so he and his friends could earn more bucks. I also got questions from the parents of Matthew's friends when they came home talking about doing chores at Mr. Justice's house.

APPENDIX E

Suggested Reading

The Sky's the Limit – Honeywell's Space and Strategic Systems Operation Blasts Off—PM Magazine, Vol 14, #4, page 16. © Aubrey Daniels International. Used with permission. https://www.aubreydaniels.com/media-center/archive/performance-management-magazine-fall-1996-vol-14-no-4

MIBE – Make International Business Easy—PM Magazine, Vol 8, #4, page 19. © Aubrey Daniels International. Used with permission. https://www.aubreydaniels.com/media-center/archive/performance-management-magazine-fall-winter-1990-vol-8-no-4

Making International Business Easy (MIBE) —PM Magazine, Vol 9, #4, page 10. © Aubrey Daniels International. Used with permission. https://www.aubreydaniels.com/media-center/archive/performance-management-magazine-fall-winter-1991-vol-9-no-4

"That Won't Work Here": Positive Reinforcement in Mexico—PM Magazine, Vol 15, #2, page 3. © Aubrey Daniels International. Used with permission. https://www.aubreydaniels.com/media-center/archive/performance-management-magazine-spring-1997-vol-15-no-2

Super Service Person Saves the Customer—PM Magazine, Vol 9, #1, page 3. © Aubrey Daniels International. Used with permission. https://www.aubreydaniels.com/media-center/archive/performance-management-magazine-winter-spring-1991-vol-9-no-1

Eastman Kodak's Chemical Division "Develops" Quality Performance. PM Magazine Vol 3, #3, page 30. © Aubrey Daniels International. Used with permission. https://www.aubreydaniels.com/media-center/archive/performance-management-magazine-spring-summer-1985-vol-3-no-3

PM and the Quality Revolution – An Interview with Russell Justice – PM Magazine, Vol 9, #2, page 12. © Aubrey Daniels International. Used with permission. https://www.aubreydaniels.com/media-center/archive/performance-management-magazine-spring-summer-1991-vol-9-no-2

Eastman's Winning Ways, PM Magazine, Vol 13, #4, page 3. © Aubrey Daniels International. Used with permission. https://www.aubreydaniels.com/media-center/archive/performance-management-magazine-fall-1995-vol-13-no-4

Performance Management Goes to School, PM Magazine, Vol 7, #2, page 18. © Aubrey Daniels International. Used with permission. https://www.aubreydaniels.com/media-center/archive/performance-management-magazine-springsummer-1989-vol-7-no-2

Performance Management in a Marketing Area, PM Magazine, Vol 5, #1, page 14. © Aubrey Daniels International. Used with permission. https://www. aubreydaniels.com/media-center/archive/performance-management-magazine -fallwinter-1986-vol-5-no-1

From Chaos to Harmony on the Boys' Ranch, PM Magazine, Vol 5, #1, page 18. © Aubrey Daniels International. Used with permission. https://www. aubreydaniels.com/media-center/archive/performance-management-magazine -fallwinter-1986-vol-5-no-1

Russell Justice and His Amazing Performance Management Machine, PM Magazine, Vol 4, #4, page 26. © Aubrey Daniels International. Used with permission. https://www.aubreydaniels.com/media-center/archive/performance -management-magazine-summerfall-1986-vol-4-no-4

McClaskey Excellence Institute. (2010, February 24). *American Greeting wraps up success with PAL's BEI*. https://mcclaskeyexcellence.com/american-greeting -wraps-up-success-with-pals-bei/

References

Allen, J. (1990). *I saw what you did and I know who you are: Bloopers, blunders and success stories on giving and receiving recognition*. Performance Management Publications.

Barter Theatre. (n.d.). *Barter Theatre history: A glimpse at our beginning*. https://bartertheatre.com/history/?toggle=A-Brief-History

Chick-fil-A. (2015, September 8). *Truett Cathy was fond of saying, "How do you know if someone needs encouragement? If they are breathing." Join us* [Image attached] [Status update]. Facebook. https://www.facebook.com/photo?fbid=10153849973195101&set=a.10151547409785101

Collins, J. (2023, March 30). *Career advancement newsletter: From critic to encourager!* LinkedIn. https://www.linkedin.com/pulse/from-critic-encourager-jimmy-collins

Daniels, A. C. (2000). *Bringing out the best in people: How to apply the astonishing power of positive reinforcement*. McGraw-Hill.

Daniels, A. C. (2016). *Bringing out the best in people: How to apply the astonishing power of positive reinforcement* (3rd ed.). McGraw Hill.

Daniels, A. C., & Rosen, T. A. (1989). *Performance management: Improving quality and productivity through positive reinforcement*. Performance Management Publications.

Deming, W. E. (2018). *Out of the crisis*. Massachusetts Institute of Technology.

From chaos to harmony on the boys' ranch. *Performance Management Magazine, 5*(1), 18–19. © Aubrey Daniels International. Used with permission. https://www.aubreydaniels.com/media-center/archive/performance-management-magazine-fallwinter-1986-vol-5-no-1

Imai, M. (2012). *Gemba kaizen: A commonsense approach to a continuous improvement strategy* (2nd ed.). McGraw Hill.

LeBoeuf, M. (1985). *The greatest management principle in the world*. Putnam.

Making international business easy (MIBE). *Performance Management Magazine, 9*(4), 10. © Aubrey Daniels International. Used with permission. https://www.aubreydaniels.com/media-center/archive/performance-management-magazine-fall-winter-1991-vol-9-no-4

Maxwell, J. C. [@johncmaxwell]. (2023, June 23). *Always remember this: Your talk talks and your walk talks, but your walk talks louder than your talk talks!* Instagram. https://www.instagram.com/p/Ct1-SbfOhlk/?img_index=1

McAlister, A. (n.d.). *About*. HIStory. https://historycloth.com/about/

McClaskey Excellence Institute. (2010, February 24). *American Greeting wraps up success with PAL's BEI.* https://mcclaskeyexcellence.com/american-greeting -wraps-up-success-with-pals-bei/

MIBE: Make international business easy. *Performance Management Magazine, 8*(4), 19. https://www.aubreydaniels.com/media-center/archive/performance -management-magazine-fall-winter-1990-vol-8-no-4

Peters, T. J., & Waterman, Jr., R. H. (1982). *In search of excellence: Lessons from America's best-run companies.* Harper & Row.

"That won't work here": Positive reinforcement in Mexico. *Performance Management Magazine, 15*(2), 3. © Aubrey Daniels International. Used with permission. https://www.aubreydaniels.com/media-center/archive/performance -management-magazine-spring-1997-vol-15-no-2

The sky's the limit – Honeywell's Space and Strategic Systems Operation blasts off. *Performance Management Magazine, 14*(4), 16. © Aubrey Daniels International. Used with permission. https://www.aubreydaniels.com/media-center/archive/ performance-management-magazine-fall-1996-vol-14-no-4

Walton, S. (1992). *Made in America: My story.* Bantam Books.

About the Author

Russell E. Justice, P.E.

Russell has over 5 decades of performance improvement experience in more than 20 countries in Latin America, North America, South America, the Asia/Pacific region, and Europe. He has consulted in the chemicals, space/aviation, printing, electronics, textiles, mining, sporting goods, heavy equipment, food services, and pulp/paper industries and in education, publishing, healthcare, theater, engineering and construction, government, and utilities. He has worked to accelerate improvement by designing, deploying, and coaching from the frontlines to the offices of CEOs at more than 50 organizations including American Greetings, Honeywell Aerospace, Alcoa, Lowe's, Pal's Sudden Service, Mead Paper, Mission Hospital, Wilson Sporting Goods, Marriott, and Barter Theatre.

He is recognized in the quality management community as a pioneer in the worldwide implementation of a quality of management process that fully integrates the key aspects of both Total Quality Management (TQM) and applied behavior analysis (ABA). He retired from Eastman Chemical Company, where he served as a senior technical associate. At Eastman, he worked throughout the company at all levels, in all organizations, and in all regions of the world. His assignments included project-improvement work, engineering supervisor, fuels coordinator, management technology development, and consulting assignments. Russell studied under, interacted with, and was mentored by thought leaders in both the TQM and ABA arenas including Aubrey Daniels, Edwards Deming, Joseph Juran, and Brian Joiner.

Russell was cofounder of The Transformation Network and has served with the National Center for Quality, the American Society for Quality, and the Institute of Industrial and Systems Engineers as workshop leader. He has served as a lead instructor/consultant for the Business Excellence Institute.

He is a popular speaker and consultant for community and professional organizations and participates in numerous off-the-job improvement efforts at church, home, and school and in the community.

NEXT STEPS—AVAILABLE FROM THE AUTHOR

RussellJustice.com

- **"This Is What Leaders Do" Seminar** *(2-hour overview)*

- **Accelerating Continuous Improvement—The Workshop Experience** *(3-day workshop for leadership teams)*

- **Instructional Materials for Business Excellence** *(1/2- to 2-hour tutorials)*

  - *Building Excellence in Performance*
    - Beyond Thank You—The Art and Science of Positive Reinforcement
    - Feedback is the Breakfast of Champions—Transforming Measures Into Feedback and Designing Creative Scoreboards That Pass the Walk-By Test
    - ABCs of Working With People—It's Not as Hard as We Make It
    - Beyond Projects to Enterprise-Wide Improvement—Mobilizing and Unifying the Workforce
    - Maximizing the Yield of Your Training Process

  - *Building Excellence in Leadership*
    - In Search of Commitment—Guaranteed Buy-In
    - Meetings That Succeed
    - It's Amazing What Praising Can Do—What to reinforce. Who to reinforce. When to reinforce. How to reinforce.
    - Quality Celebrations—How to Celebrate Improvement

  - *Building Excellence in Customer Relationships*
    - Customer Delight—4 Simple, Proven Tools for Immediate Impact
    - Exec-to-Exec Customer Stewardship
    - Hearing and Responding to the Voice of the Customer

  - *Personal Development*
    - Whine, Bark, Growl—Dealing With Difficult People
    - Getting Help From Others
    - Converting "Ought To's" Into Actions
    - The Key to Emotional Relationships—The Five Love Languages
    - Build a Stage for Them to Stand On—Recognizing and Reinforcing Personal Differences

- **Speaker Engagements** *(for conferences, trade shows, professional organizations, leadership retreats, and marketing/sales meetings)*

RELATED READING AVAILABLE FROM
KEYPRESS PUBLISHING

KeyPressPublishing.com

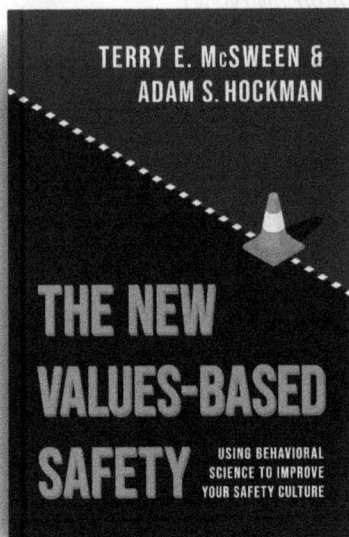

The New Values-Based Safety
Terry McSween & Adam Hockman

The New Values-Based Safety is an essential reference for leaders and safety professionals in any industry. The book covers safety process implementation from design to execution and maintenance, plus special considerations various companies's needs, the behavioral science behind it all, and real-world case studies. Written in understandable language, the book underscores the importance of creating a culture of caring and concern to support employee well-being along with the company's bottom line.

valuesbasedsafety.com

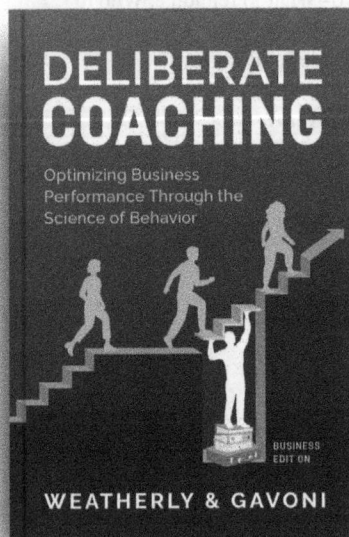

Deliberate Coaching (Business Edition)
Nicholas L. Weatherly & Paul Gavoni

Deliberate Coaching transforms organizations with tools enhancing leader effectiveness and organizational performance. Unlike traditional coaching guides, this book delivers a science-based approach with proven and practical strategies for lasting positive change. It addresses shortcomings of conventional training and performance-improvement methods and offers alternatives that can be applied quickly and easily. *Deliberate Coaching* clarifies the proper use and role of behavioral consequences and advocates for a precise, purposeful, and systematic coaching strategy.

deliberatecoachingbiz.com